ACT Aspire Subject Test Mathematics Grade 7

Student Practice Workbook

+ Two Full-Length ACT Aspire Math Tests

Math Notion

www.MathNotion.com

ACT Aspire Subject Test Mathematics Grade 7

ACT Aspire Subject Test Mathematics Grade 7

ACT Aspire Subject Test Mathematics Grade 7

Published in the United State of America By

The Math Notion

Web: WWW.MathNotion.com

Email: info@Mathnotion.com

Copyright © 2021 by the Math Notion. All rights reserved. No part of this publication may be reproduced, stored in a retrieval system, or transmitted in any form or by any means, electronic, mechanical, photocopying, recording, scanning, or otherwise, except as permitted under Section 107 or 108 of the 1976 United States Copyright Ac, without permission of the author.

All inquiries should be addressed to the Math Notion.

ISBN: 978-1-63620-068-2

ACT Aspire Subject Test Mathematics Grade 7

The Math Notion

Michael Smith has been a math instructor for over a decade now. He launched the Math Notion. Since 2006, we have devoted our time to both teaching and developing exceptional math learning materials. As a test prep company, we have worked with thousands of students. We have used the feedback of our students to develop a unique study program that can be used by students to drastically improve their math scores fast and effectively. We have more than a thousand Math learning books including:

– SAT Math Prep

– ACT Math Prep

– SSAT/ISEE Math Prep

– Accuplacer Math Prep

– ACT Aspire Math Prep

–many Math Education Workbooks, Study Guides, Practice and Exercise Books

As an experienced Math test preparation company, we have helped many students raise their standardized test scores—and attend the colleges of their dreams. We tutor online and in person, we teach students in large groups, and we provide training materials and textbooks through our website and through Amazon.

You can contact us via email at:

info@Mathnotion.com

ACT Aspire Subject Test Mathematics Grade 7

Get the Targeted Practice You Need to Ace the ACT Aspire Math Test!

ACT Aspire Subject Test Mathematics Grade 7 includes easy-to-follow instructions, helpful examples, and plenty of math practice problems to assist students to master each concept, brush up their problem-solving skills, and create confidence.

The ACT Aspire math practice book provides numerous opportunities to evaluate basic skills along with abundant remediation and intervention activities. It is a skill that permits you to quickly master intricate information and produce better leads in less time.

Students can boost their test-taking skills by taking the book's two practice ACT Aspire Math exams. All test questions answered and explained in detail.

Important Features of the 7th grade ACT Aspire Math Book:

- A **complete review** of ACT Aspire math test topics,
- Over 2,500 practice problems covering all topics tested,
- The most important concepts you need to know,
- Clear and concise, easy-to-follow sections,
- Well designed for enhanced learning and interest,
- Hands-on experience with all question types,
- **2 full-length practice tests** with detailed answer explanations,
- Cost-Effective Pricing,

Powerful math exercises to help you avoid traps and pacing yourself to beat the ACT Aspire test. Students will gain valuable experience and raise their confidence by taking 7th grade math practice tests, learning about test structure, and gaining a deeper understanding of what is tested on the ACT Aspire math grade 7. If ever there was a book to respond to the pressure to increase students' test scores, this is it.

WWW.MathNotion.COM

… So Much More Online!

✓ FREE Math Lessons

✓ More Math Learning Books!

✓ Mathematics Worksheets

✓ Online Math Tutors

For a PDF Version of This Book

Please Visit WWW.MathNotion.com

ACT Aspire Subject Test Mathematics Grade 7

Contents

Chapter 1 : Integers and Number Theory **11**
   Rounding 12
   Rounding and Estimates 13
   Adding and Subtracting Integers 14
   Multiplying and Dividing Integers 15
   Order of Operations 16
   Ordering Integers and Numbers 17
   Integers and Absolute Value 18
   Factoring Numbers 19
   Greatest Common Factor 20
   Least Common Multiple 21
   Answers of Worksheets 22

Chapter 2 : Fractions **25**
   Simplifying Fractions 26
   Adding and Subtracting Fractions 27
   Multiplying and Dividing Fractions 28
   Adding and Subtracting Mixed Numbers 29
   Multiplying and Dividing Mixed Numbers 30
   Answers of Worksheets 31

Chapter 3 : Decimals **33**
   Adding and Subtracting Decimals 34
   Multiplying and Dividing Decimals 35
   Comparing Decimals 36
   Rounding Decimals 37
   Answers of Worksheets 38

Chapter 4 : Proportions, Ratios, and Percent **39**
   Simplifying Ratios 40
   Proportional Ratios 41
   Similarity and Ratios 42
   Ratio and Rates Word Problems 43
   Percentage Calculations 44
   Percent Problems 45
   Discount, Tax and Tip 46

WWW.MathNotion.Com

ACT Aspire Subject Test Mathematics Grade 7

Percent of Change 47
Simple Interest 48
Answers of Worksheets 49

Chapter 5 : Exponents and Radicals Expressions 52
Multiplication Property of Exponents 53
Zero and Negative Exponents 54
Division Property of Exponents 55
Powers of Products and Quotients 56
Negative Exponents and Negative Bases 57
Scientific Notation 58
Square Roots 59
Answers of Worksheets 60

Chapter 6 : Algebraic Expressions 63
Translate Phrases into an Algebraic Statement 64
Simplifying Variable Expressions 65
The Distributive Property 66
Evaluating One Variable Expressions 67
Evaluating Two Variables Expressions 68
Combining like Terms 69
Answers of Worksheets 70

Chapter 7 : Equations and Inequalities 72
One–Step Equations 73
Multi–Step Equations 74
Graphing Single–Variable Inequalities 75
One–Step Inequalities 76
Multi-Step Inequalities 77
Systems of Equations 78
Systems of Equations Word Problems 79
Answers of Worksheets 80

Chapter 8 : Linear Functions 83
Finding Slope 84
Graphing Lines Using Line Equation 85
Writing Linear Equations 86
Graphing Linear Inequalities 87
Finding Midpoint 88
Finding Distance of Two Points 89
Answers of Worksheets 90

ACT Aspire Subject Test Mathematics Grade 7

Chapter 9 : Transformations 93
    Translations 94
    Reflections 95
    Rotations 97
    Dilations 99
    Coordinates of Vertices 100
    Answers of Worksheets 101

Chapter 10 : Geometry and Solid Figures 104
    Angles 105
    Pythagorean Relationship 106
    Triangles 107
    Polygons 108
    Trapezoids 109
    Circles 110
    Cubes 111
    Rectangular Prism 111
    Cylinder 113
    Pyramids and Cone 114
    Answers of Worksheets 115

Chapter 11 : Statistics and Probability 117
    Mean and Median 118
    Mode and Range 119
    Times Series 120
    Stem-and-Leaf Plot 121
    Pie Graph 122
    Probability Problems 123
    Answers of Worksheets 124

Chapter 12 : ACT Aspire Math Practice Tests 127
    ACT Aspire GRADE 7 MAHEMATICS REFRENCE MATERIALS 129
    ACT Aspire Practice Test 1 131
    ACT Aspire Practice Test 2 147

Answers and Explanations 161
    Answer Key 161
    Practice Test 1 163
    Practice Test 2 170

ACT Aspire Subject Test Mathematics Grade 7

Chapter 1 :
Integers and Number Theory

Topics that you will practice in this chapter:

- ✓ Rounding
- ✓ Rounding and Estimates
- ✓ Adding and Subtracting Integers
- ✓ Multiplying and Dividing Integers
- ✓ Order of Operations
- ✓ Ordering Integers and Numbers
- ✓ Integers and Absolute Value
- ✓ Factoring Numbers
- ✓ Greatest Common Factor (GCF)
- ✓ Least Common Multiple (LCM)

"Wherever there is number, there is beauty." −Proclus

ACT Aspire Subject Test Mathematics Grade 7

Rounding

✎ **Round each number to the nearest ten.**

1) 42 = ___
2) 88 = ___
3) 24 = ___
4) 57 = ___

5) 19 = ___
6) 25 = ___
7) 93 = ___
8) 71 = ___

9) 48 = ___
10) 81 = ___
11) 58 = ___
12) 87 = ___

✎ **Round each number to the nearest hundred.**

13) 198 = ___
14) 387 = ___
15) 816 = ___
16) 101 = ___

17) 321 = ___
18) 433 = ___
19) 579 = ___
20) 825 = ___

21) 580 = ___
22) 868 = ___
23) 480 = ___
24) 287 = ___

✎ **Round each number to the nearest thousand.**

25) 1,382 = ___
26) 3,420 = ___
27) 4,254 = ___
28) 6,861 = ___

29) 9,099 = ___
30) 22,980 = ___
31) 45,188 = ___
32) 16,808 = ___

33) 52,866 = ___
34) 85,190 = ___
35) 70,990 = ___
36) 26,869 = ___

WWW.MathNotion.Com

ACT Aspire Subject Test Mathematics Grade 7

Rounding and Estimates

✎ **Estimate the sum by rounding each number to the nearest ten.**

1) $13 + 22 =$ _____

2) $71 + 23 =$ _____

3) $61 + 58 =$ _____

4) $56 + 85 =$ _____

5) $368 + 249 =$ _____

6) $330 + 903 =$ _____

7) $471 + 293 =$ _____

8) $1,950 + 2,655 =$ _____

✎ **Estimate the product by rounding each number to the nearest ten.**

9) $32 \times 71 =$ _____

10) $12 \times 33 =$ _____

11) $31 \times 83 =$ _____

12) $19 \times 11 =$ _____

13) $42 \times 76 =$ _____

14) $63 \times 34 =$ _____

15) $19 \times 31 =$ _____

16) $59 \times 71 =$ _____

✎ **Estimate the sum or product by rounding each number to the nearest ten.**

17) $\begin{array}{r} 29 \\ \times\ 12 \\ \hline \end{array}$

18) $\begin{array}{r} 37 \\ \times\ 26 \\ \hline \end{array}$

19) $\begin{array}{r} 48 \\ +\ 82 \\ \hline \end{array}$

20) $\begin{array}{r} 65 \\ +44 \\ \hline \end{array}$

21) $\begin{array}{r} 37 \\ \times\ 14 \\ \hline \end{array}$

22) $\begin{array}{r} 71 \\ +\ 32 \\ \hline \end{array}$

WWW.MathNotion.Com

ACT Aspire Subject Test Mathematics Grade 7

Adding and Subtracting Integers

✏ **Find each sum.**

1) $14 + (-6) =$

2) $(-13) + (-20) =$

3) $5 + (-28) =$

4) $50 + (-12) =$

5) $(-7) + (-15) + 3 =$

6) $30 + (-14) + 8 =$

7) $40 + (-10) + (-14) + 17 =$

8) $(-15) + (-20) + 13 + 35 =$

9) $40 + (-20) + (38 - 29) =$

10) $28 + (-12) + (30 - 12) =$

✏ **Find each difference.**

11) $(-18) - (-7) =$

12) $25 - (-14) =$

13) $(-20) - 36 =$

14) $34 - (-19) =$

15) $51 - (30 - 21) =$

16) $17 - (5) - (-24) =$

17) $(35 + 20) - (-46) =$

18) $48 - 16 - (-8) =$

19) $62 - (28 + 17) - (-15) =$

20) $58 - (-23) - (-31) =$

21) $19 - (-8) - (-13) =$

22) $(19 - 24) - (-14) =$

23) $27 - 33 - (-21) =$

24) $58 - (32 + 24) - (-9) =$

25) $36 - (-30) + (-17) =$

26) $27 - (-42) + (-31) =$

WWW.MathNotion.Com

ACT Aspire Subject Test Mathematics Grade 7

Multiplying and Dividing Integers

✎ **Find each product.**

1) $(-9) \times (-5) =$

2) $(-3) \times 9 =$

3) $8 \times (-12) =$

4) $(-7) \times (-20) =$

5) $(-3) \times (-5) \times 6 =$

6) $(14 - 3) \times (-8) =$

7) $12 \times (-9) \times (-3) =$

8) $(140 + 10) \times (-2) =$

9) $10 \times (-12 + 8) \times 3 =$

10) $(-8) \times (-5) \times (-10) =$

✎ **Find each quotient.**

11) $42 \div (-7) =$

12) $(-48) \div (-6) =$

13) $(-40) \div (-8) =$

14) $54 \div (-2) =$

15) $152 \div 19 =$

16) $(-144) \div (-12) =$

17) $180 \div (-10) =$

18) $(-312) \div (-12) =$

19) $221 \div (-13) =$

20) $(-126) \div (6) =$

21) $(-161) \div (-7) =$

22) $-266 \div (-14) =$

23) $(-120) \div (-4) =$

24) $270 \div (-18) =$

25) $(-208) \div (-8) =$

26) $(135) \div (-15) =$

ACT Aspire Subject Test Mathematics Grade 7

Order of Operations

✏ **Evaluate each expression.**

1) $7 + (5 \times 4) =$

2) $14 - (3 \times 6) =$

3) $(19 \times 4) + 16 =$

4) $(16 - 7) - (8 \times 2) =$

5) $27 + (18 \div 3) =$

6) $(18 \times 8) \div 6 =$

7) $(32 \div 4) \times (-2) =$

8) $(9 \times 4) + (32 - 18) =$

9) $24 + (4 \times 3) + 7 =$

10) $(36 \times 3) \div (2 + 2) =$

11) $(-7) + (12 \times 3) + 11 =$

12) $(8 \times 5) - (24 \div 6) =$

13) $(7 \times 6 \div 3) - (12 + 9) =$

14) $(13 + 5 - 14) \times 3 - 2 =$

15) $(20 - 14 + 30) \times (64 \div 4) =$

16) $32 + (28 - (36 \div 9)) =$

17) $(7 + 6 - 4 - 7) + (15 \div 5) =$

18) $(85 - 20) + (20 - 18 + 7) =$

19) $(20 \times 2) + (14 \times 3) - 22 =$

20) $18 + 5 - (30 \times 3) + 20 =$

ACT Aspire Subject Test Mathematics Grade 7

Ordering Integers and Numbers

✎ **Order each set of integers from least to greatest.**

1) $8, -10, -5, -3, 4$ ___, ___, ___, ___, ___, ___

2) $-10, -18, 6, 14, 27$ ___, ___, ___, ___, ___, ___

3) $15, -8, -21, 21, -23$ ___, ___, ___, ___, ___, ___

4) $-14, -40, 23, -12, 47$ ___, ___, ___, ___, ___, ___

5) $59, -54, 32, -57, 36$ ___, ___, ___, ___, ___, ___

6) $68, 26, -19, 47, -34$ ___, ___, ___, ___, ___, ___

✎ **Order each set of integers from greatest to least.**

7) $18, 36, -16, -18, -10$ ___, ___, ___, ___, ___, ___

8) $27, 34, -12, -24, 94$ ___, ___, ___, ___, ___, ___

9) $50, -21, -13, 42, -2$ ___, ___, ___, ___, ___, ___

10) $37, 46, -20, -16, 86$ ___, ___, ___, ___, ___, ___

11) $-18, 88, -26, -59, 75$ ___, ___, ___, ___, ___, ___

12) $-65, -30, -25, 3, 14$ ___, ___, ___, ___, ___, ___

ACT Aspire Subject Test Mathematics Grade 7

Integers and Absolute Value

✏️ **Write absolute value of each number.**

1) $|-2| =$

2) $|-27| =$

3) $|-20| =$

4) $|14| =$

5) $|6| =$

6) $|-55| =$

7) $|16| =$

8) $|2| =$

9) $|54| =$

10) $|-4| =$

11) $|-11| =$

12) $|88| =$

13) $|0| =$

14) $|79| =$

15) $|-32| =$

16) $|-17| =$

17) $|42| =$

18) $|-46| =$

19) $|1| =$

20) $|-40| =$

✏️ **Evaluate the value.**

21) $|-5| - \dfrac{|-21|}{7} =$

22) $14 - |3 - 15| - |-4| =$

23) $\dfrac{|-32|}{4} \times |-4| =$

24) $\dfrac{|7 \times (-3)|}{7} \times \dfrac{|-19|}{3} =$

25) $|4 \times (-5)| + \dfrac{|-40|}{5} =$

26) $\dfrac{|-45|}{9} \times \dfrac{|-24|}{12} =$

27) $|-12 + 8| \times \dfrac{|-7 \times 7|}{7} =$

28) $\dfrac{|-11 \times 2|}{4} \times |-16| =$

WWW.MathNotion.Com

Factoring Numbers

✎ **List all positive factors of each number.**

1) 9

2) 16

3) 24

4) 30

5) 26

6) 46

7) 20

8) 68

9) 28

10) 98

11) 14

12) 54

13) 55

14) 18

15) 63

16) 34

17) 50

18) 62

19) 95

20) 64

21) 70

22) 45

23) 22

24) 65

Greatest Common Factor

👉 **Find the GCF for each number pair.**

1) 6, 2

2) 4, 5

3) 3, 12

4) 7, 3

5) 5, 10

6) 8, 48

7) 6, 18

8) 9, 15

9) 12, 18

10) 4, 36

11) 6, 10

12) 28, 52

13) 25, 10

14) 22, 24

15) 9, 54

16) 8, 54

17) 42, 14

18) 16, 40

19) 9, 2, 3

20) 5, 15, 10

21) 7, 9, 2

22) 16, 64

23) 30, 48

24) 36, 63

ACT Aspire Subject Test Mathematics Grade 7

Least Common Multiple

✎ **Find the LCM for each number pair.**

1) 6, 9

2) 15, 45

3) 16, 40

4) 12, 36

5) 18, 27

6) 14, 42

7) 6, 30

8) 8, 56

9) 7, 21

10) 8, 20

11) 15, 25

12) 7, 9

13) 4, 11

14) 8, 28

15) 28, 56

16) 40, 50

17) 12, 13

18) 22, 11

19) 36, 20

20) 15, 35

21) 18, 81

22) 30, 54

23) 18, 45

24) 75, 25

ACT Aspire Subject Test Mathematics Grade 7

Answers of Worksheets

Rounding

1) 40	10) 80	19) 600	28) 7,000
2) 90	11) 60	20) 800	29) 9,000
3) 20	12) 90	21) 600	30) 23,000
4) 60	13) 200	22) 900	31) 45,000
5) 20	14) 400	23) 500	32) 17,000
6) 30	15) 800	24) 300	33) 53,000
7) 90	16) 100	25) 1,000	34) 85,000
8) 70	17) 300	26) 3,000	35) 71,000
9) 50	18) 400	27) 4,000	36) 27,000

Rounding and Estimates

1) 30	7) 760	13) 3,200	19) 130
2) 90	8) 4,610	14) 1,800	20) 110
3) 120	9) 2,100	15) 600	21) 400
4) 150	10) 300	16) 4,200	22) 100
5) 620	11) 2,400	17) 300	
6) 1,230	12) 200	18) 1,200	

Adding and Subtracting Integers

1) 8	8) 13	15) 42	22) 9
2) −33	9) 29	16) 36	23) 15
3) −23	10) 34	17) 101	24) 11
4) 38	11) −11	18) 40	25) 49
5) −19	12) 39	19) 32	26) 38
6) 24	13) −56	20) 112	
7) 33	14) 53	21) 40	

Multiplying and Dividing Integers

1) 45	6) −88	11) −6	16) 12
2) −27	7) 324	12) 8	17) −18
3) −96	8) −300	13) 5	18) 26
4) 140	9) −120	14) −27	19) −17
5) 90	10) −400	15) 8	20) −21

WWW.MathNotion.Com

ACT Aspire Subject Test Mathematics Grade 7

21) 23
22) 19
23) 30
24) −15
25) 26
26) −9

Order of Operations

1) 27
2) −4
3) 92
4) −7
5) 33
6) 24
7) −16
8) 50
9) 43
10) 27
11) 40
12) 36
13) −7
14) 10
15) 576
16) 56
17) 5
18) 74
19) 60
20) −47

Ordering Integers and Numbers

1) −10, −5, −3, 4, 8
2) −18, −10, 6, 14, 27
3) −23, −21, −8, 15, 21
4) −40, −14, −12, 23, 47
5) −57, −54, 32, 36, 59
6) −34, −19, 26, 47, 68
7) 36, 18, −10, −16, −18
8) 94, 34, 27, −12, −24
9) 50, 42, −2, −13, −21
10) 86, 46, 37, −16, −20
11) 88, 75, −18, −26, −59
12) 14, 3, −25, −30, −65

Integers and Absolute Value

1) 2
2) 27
3) 20
4) 14
5) 6
6) 55
7) 16
8) 2
9) 54
10) 4
11) 11
12) 88
13) 0
14) 79
15) 32
16) 17
17) 42
18) 46
19) 1
20) 40
21) 2
22) −2
23) 32
24) 19
25) 28
26) 10
27) 28
28) 88

Factoring Numbers

1) 1, 3, 9
2) 1, 2, 4, 8, 16
3) 1, 2, 3, 4, 6, 8, 12, 24
4) 1, 2, 3, 5, 6, 10, 15, 30
5) 1, 2, 13, 26
6) 1, 2, 23, 46
7) 1, 2, 4, 5, 10, 20
8) 1, 2, 4, 17, 34, 68
9) 1, 2, 4, 7, 14, 28
10) 1, 2, 7, 14, 49, 98
11) 1, 2, 7, 14
12) 1, 2, 3, 6, 9, 18, 27, 54
13) 1, 5, 11, 55
14) 1, 2, 3, 6, 9, 18
15) 1, 3, 7, 9, 21, 63
16) 1, 2, 17, 34
17) 1, 2, 5, 10, 25, 50
18) 1, 2, 31, 62
19) 1, 5, 19, 95
20) 1, 2, 4, 8, 16, 32, 64
21) 1, 2, 5, 7, 10, 14, 35, 70
22) 1, 3, 5, 9, 15, 45
23) 1, 2, 11, 22
24) 1, 5, 13, 65

WWW.MathNotion.Com

ACT Aspire Subject Test Mathematics Grade 7

Greatest Common Factor

1) 2
2) 1
3) 3
4) 1
5) 5
6) 8
7) 6
8) 3
9) 6
10) 4
11) 2
12) 4
13) 5
14) 2
15) 9
16) 2
17) 14
18) 8
19) 1
20) 5
21) 1
22) 16
23) 6
24) 9

Least Common Multiple

1) 18
2) 45
3) 80
4) 36
5) 54
6) 42
7) 30
8) 56
9) 21
10) 40
11) 75
12) 63
13) 44
14) 56
15) 56
16) 200
17) 156
18) 22
19) 180
20) 105
21) 162
22) 270
23) 90
24) 75

ACT Aspire Subject Test Mathematics Grade 7

Chapter 2 :
Fractions

Topics that you will practice in this chapter:

- ✓ Simplifying Fractions
- ✓ Adding and Subtracting Fractions
- ✓ Multiplying and Dividing Fractions
- ✓ Adding and Subtract Mixed Numbers
- ✓ Multiplying and Dividing Mixed Numbers

"A Man is like a fraction whose numerator is what he is and whose denominator is what he thinks of himself. The larger the denominator, the smaller the fraction." –Tolstoy

ACT Aspire Subject Test Mathematics Grade 7

Simplifying Fractions

✎ **Simplify each fraction to its lowest terms.**

1) $\dfrac{5}{10} =$

2) $\dfrac{28}{35} =$

3) $\dfrac{27}{36} =$

4) $\dfrac{40}{80} =$

5) $\dfrac{14}{56} =$

6) $\dfrac{32}{48} =$

7) $\dfrac{52}{65} =$

8) $\dfrac{15}{60} =$

9) $\dfrac{80}{160} =$

10) $\dfrac{55}{77} =$

11) $\dfrac{28}{112} =$

12) $\dfrac{32}{64} =$

13) $\dfrac{63}{72} =$

14) $\dfrac{81}{90} =$

15) $\dfrac{35}{105} =$

16) $\dfrac{25}{70} =$

17) $\dfrac{80}{280} =$

18) $\dfrac{12}{81} =$

19) $\dfrac{36}{186} =$

20) $\dfrac{240}{540} =$

21) $\dfrac{70}{560} =$

✎ **Find the answer for each problem.**

22) Which of the following fractions equal to $\dfrac{3}{4}$? ____

　A. $\dfrac{60}{90}$　　B. $\dfrac{43}{104}$　　C. $\dfrac{48}{64}$　　D. $\dfrac{150}{300}$

23) Which of the following fractions equal to $\dfrac{5}{8}$? ____

　A. $\dfrac{125}{200}$　　B. $\dfrac{115}{200}$　　C. $\dfrac{50}{100}$　　D. $\dfrac{30}{90}$

24) Which of the following fractions equal to $\dfrac{3}{7}$? ____

　A. $\dfrac{58}{116}$　　B. $\dfrac{54}{126}$　　C. $\dfrac{270}{167}$　　D. $\dfrac{42}{63}$

WWW.MathNotion.Com

ACT Aspire Subject Test Mathematics Grade 7

Adding and Subtracting Fractions

✎ Find the sum.

1) $\frac{5}{9} + \frac{4}{9} =$

2) $\frac{1}{2} + \frac{1}{7} =$

3) $\frac{3}{8} + \frac{1}{4} =$

4) $\frac{3}{5} + \frac{1}{2} =$

5) $\frac{1}{4} + \frac{3}{5} =$

6) $\frac{7}{8} + \frac{3}{8} =$

7) $\frac{1}{2} + \frac{7}{10} =$

8) $\frac{2}{5} + \frac{2}{3} =$

9) $\frac{5}{7} + \frac{2}{3} =$

10) $\frac{7}{12} + \frac{3}{4} =$

11) $\frac{5}{6} + \frac{2}{5} =$

12) $\frac{1}{12} + \frac{2}{3} =$

✎ Find the difference.

13) $\frac{1}{3} - \frac{1}{6} =$

14) $\frac{3}{4} - \frac{1}{8} =$

15) $\frac{1}{2} - \frac{1}{3} =$

16) $\frac{1}{4} - \frac{1}{5} =$

17) $\frac{5}{8} - \frac{2}{3} =$

18) $\frac{1}{4} - \frac{1}{7} =$

19) $\frac{5}{6} - \frac{1}{9} =$

20) $\frac{3}{4} - \frac{1}{6} =$

21) $\frac{7}{8} - \frac{1}{12} =$

22) $\frac{8}{15} - \frac{3}{5} =$

23) $\frac{3}{12} - \frac{1}{14} =$

24) $\frac{10}{13} - \frac{7}{26} =$

25) $\frac{6}{7} - \frac{3}{4} =$

26) $\frac{4}{5} - \frac{1}{8} =$

27) $\frac{4}{7} - \frac{2}{35} =$

28) $\frac{9}{16} - \frac{2}{8} =$

29) $\frac{8}{9} - \frac{7}{18} =$

30) $\frac{1}{2} - \frac{4}{9} =$

WWW.MathNotion.Com

ACT Aspire Subject Test Mathematics Grade 7

Multiplying and Dividing Fractions

✎ Find the value of each expression in lowest terms.

1) $\dfrac{1}{5} \times \dfrac{15}{5} =$

2) $\dfrac{9}{12} \times \dfrac{4}{9} =$

3) $\dfrac{1}{16} \times \dfrac{8}{10} =$

4) $\dfrac{1}{24} \times \dfrac{8}{10} =$

5) $\dfrac{1}{5} \times \dfrac{1}{4} =$

6) $\dfrac{7}{9} \times \dfrac{1}{7} =$

7) $\dfrac{6}{7} \times \dfrac{1}{3} =$

8) $\dfrac{2}{8} \times \dfrac{2}{8} =$

9) $\dfrac{5}{8} \times \dfrac{3}{5} =$

10) $\dfrac{4}{7} \times \dfrac{1}{8} =$

11) $\dfrac{7}{15} \times \dfrac{5}{7} =$

12) $\dfrac{3}{10} \times \dfrac{5}{9} =$

✎ Find the value of each expression in lowest terms.

13) $\dfrac{1}{4} \div \dfrac{1}{8} =$

14) $\dfrac{1}{10} \div \dfrac{1}{5} =$

15) $\dfrac{3}{4} \div \dfrac{1}{5} =$

16) $\dfrac{1}{3} \div \dfrac{5}{6} =$

17) $\dfrac{1}{7} \div \dfrac{8}{42} =$

18) $\dfrac{3}{4} \div \dfrac{1}{6} =$

19) $\dfrac{2}{7} \div \dfrac{7}{13} =$

20) $\dfrac{1}{24} \div \dfrac{3}{16} =$

21) $\dfrac{7}{12} \div \dfrac{5}{6} =$

22) $\dfrac{22}{18} \div \dfrac{11}{9} =$

23) $\dfrac{9}{35} \div \dfrac{3}{7} =$

24) $\dfrac{2}{7} \div \dfrac{8}{21} =$

25) $\dfrac{1}{9} \div \dfrac{2}{5} =$

26) $\dfrac{5}{12} \div \dfrac{3}{5} =$

27) $\dfrac{3}{20} \div \dfrac{1}{6} =$

28) $\dfrac{8}{20} \div \dfrac{3}{4} =$

29) $\dfrac{5}{6} \div \dfrac{2}{9} =$

30) $\dfrac{5}{11} \div \dfrac{3}{4} =$

WWW.MathNotion.Com

ACT Aspire Subject Test Mathematics Grade 7

Adding and Subtracting Mixed Numbers

✎ **Find the sum.**

1) $3\frac{1}{3} + 2\frac{1}{6} =$

2) $4\frac{1}{2} + 3\frac{1}{2} =$

3) $3\frac{3}{8} + 1\frac{1}{8} =$

4) $2\frac{1}{4} + 2\frac{1}{3} =$

5) $3\frac{5}{6} + 2\frac{7}{12} =$

6) $5\frac{4}{15} + 3\frac{3}{5} =$

7) $2\frac{1}{3} + 4\frac{3}{7} =$

8) $3\frac{1}{2} + 4\frac{2}{5} =$

9) $5\frac{2}{5} + 6\frac{3}{7} =$

10) $8\frac{5}{16} + 6\frac{1}{12} =$

✎ **Find the difference.**

11) $3\frac{1}{4} - 1\frac{3}{4} =$

12) $6\frac{3}{5} - 4\frac{2}{5} =$

13) $4\frac{1}{3} - 3\frac{1}{9} =$

14) $7\frac{1}{7} - 5\frac{1}{2} =$

15) $5\frac{1}{3} - 2\frac{1}{12} =$

16) $8\frac{1}{5} - 4\frac{1}{3} =$

17) $9\frac{1}{4} - 6\frac{1}{8} =$

18) $11\frac{7}{15} - 8\frac{3}{5} =$

19) $14\frac{5}{6} - 11\frac{3}{5} =$

20) $18\frac{2}{7} - 14\frac{1}{5} =$

21) $9\frac{1}{3} - 4\frac{1}{4} =$

22) $6\frac{1}{8} - 4\frac{1}{16} =$

23) $19\frac{3}{8} - 15\frac{1}{3} =$

24) $11\frac{1}{9} - 8\frac{1}{8} =$

25) $17\frac{1}{7} - 11\frac{1}{5} =$

26) $16\frac{2}{9} - 9\frac{5}{7} =$

WWW.MathNotion.Com

ACT Aspire Subject Test Mathematics Grade 7

Multiplying and Dividing Mixed Numbers

✎ **Find the product.**

1) $5\frac{1}{2} \times 2\frac{1}{4} =$

2) $5\frac{1}{3} \times 4\frac{1}{3} =$

3) $5\frac{3}{4} \times 6\frac{1}{4} =$

4) $3\frac{1}{3} \times 2\frac{3}{5} =$

5) $4\frac{8}{10} \times 1\frac{1}{24} =$

6) $6\frac{2}{7} \times 1\frac{1}{11} =$

7) $8\frac{2}{3} \times 3\frac{1}{2} =$

8) $3\frac{4}{7} \times 2\frac{1}{5} =$

9) $5\frac{2}{8} \times 4\frac{1}{6} =$

10) $7\frac{3}{3} \times 1\frac{3}{8} =$

✎ **Find the quotient.**

11) $2\frac{2}{5} \div 4\frac{1}{5} =$

12) $4\frac{1}{6} \div 3\frac{1}{3} =$

13) $6\frac{1}{3} \div 1\frac{1}{2} =$

14) $7\frac{1}{10} \div 2\frac{2}{5} =$

15) $3\frac{1}{3} \div 1\frac{1}{9} =$

16) $1\frac{1}{10} \div 4\frac{1}{2} =$

17) $1\frac{3}{16} \div 5\frac{1}{4} =$

18) $4\frac{1}{3} \div 4\frac{3}{4} =$

19) $9\frac{1}{3} \div 2\frac{1}{4} =$

20) $15\frac{1}{3} \div 5\frac{1}{2} =$

21) $4\frac{1}{6} \div 1\frac{1}{5} =$

22) $1\frac{1}{18} \div 1\frac{2}{9} =$

23) $4\frac{2}{7} \div 1\frac{3}{10} =$

24) $7\frac{1}{3} \div 2\frac{2}{11} =$

25) $8\frac{2}{5} \div 1\frac{1}{6} =$

26) $9\frac{1}{3} \div 2\frac{1}{7} =$

WWW.MathNotion.Com

ACT Aspire Subject Test Mathematics Grade 7

Answers of Worksheets

Simplifying Fractions

1) $\frac{1}{2}$
2) $\frac{4}{5}$
3) $\frac{3}{4}$
4) $\frac{1}{2}$
5) $\frac{1}{4}$
6) $\frac{2}{3}$
7) $\frac{4}{5}$
8) $\frac{1}{4}$
9) $\frac{1}{2}$
10) $\frac{5}{7}$
11) $\frac{1}{4}$
12) $\frac{1}{2}$
13) $\frac{7}{8}$
14) $\frac{9}{10}$
15) $\frac{1}{3}$
16) $\frac{5}{14}$
17) $\frac{2}{7}$
18) $\frac{4}{27}$
19) $\frac{6}{31}$
20) $\frac{4}{9}$
21) $\frac{1}{8}$
22) C
23) A
24) B

Adding and Subtracting Fractions

1) $\frac{9}{9} = 1$
2) $\frac{9}{14}$
3) $\frac{5}{8}$
4) $1\frac{1}{10}$
5) $\frac{17}{20}$
6) $1\frac{1}{4}$
7) $1\frac{1}{5}$
8) $1\frac{1}{15}$
9) $1\frac{8}{21}$
10) $1\frac{1}{3}$
11) $1\frac{7}{30}$
12) $\frac{3}{4}$
13) $\frac{1}{6}$
14) $\frac{5}{8}$
15) $\frac{1}{6}$
16) $\frac{1}{20}$
17) $-\frac{1}{24}$
18) $\frac{3}{28}$
19) $\frac{13}{18}$
20) $\frac{7}{12}$
21) $\frac{19}{24}$
22) $-\frac{1}{15}$
23) $\frac{5}{28}$
24) $\frac{1}{2}$
25) $\frac{3}{28}$
26) $\frac{27}{40}$
27) $\frac{18}{35}$
28) $\frac{5}{16}$
29) $\frac{1}{2}$
30) $\frac{1}{18}$

Multiplying and Dividing Fractions

1) $\frac{3}{5}$
2) $\frac{1}{3}$
3) $\frac{1}{20}$
4) $\frac{1}{30}$
5) $\frac{1}{20}$
6) $\frac{1}{9}$
7) $\frac{2}{7}$
8) $\frac{1}{16}$
9) $\frac{3}{8}$
10) $\frac{1}{14}$
11) $\frac{1}{3}$
12) $\frac{1}{6}$
13) 2
14) $\frac{1}{2}$
15) $3\frac{3}{4}$
16) $\frac{2}{5}$

ACT Aspire Subject Test Mathematics Grade 7

17) $\frac{3}{4}$

18) $4\frac{1}{2}$

19) $\frac{26}{49}$

20) $\frac{2}{9}$

21) $\frac{7}{10}$

22) 1

23) $\frac{3}{5}$

24) $\frac{3}{4}$

25) $\frac{5}{18}$

26) $\frac{25}{36}$

27) $\frac{9}{10}$

28) $\frac{8}{15}$

29) $3\frac{3}{4}$

30) $\frac{20}{33}$

Adding and Subtracting Mixed Numbers

1) $5\frac{1}{2}$

2) 8

3) $4\frac{1}{2}$

4) $4\frac{7}{12}$

5) $6\frac{5}{12}$

6) $8\frac{13}{15}$

7) $6\frac{16}{21}$

8) $7\frac{9}{10}$

9) $11\frac{29}{35}$

10) $14\frac{19}{48}$

11) $1\frac{1}{2}$

12) $2\frac{1}{5}$

13) $1\frac{2}{9}$

14) $1\frac{9}{14}$

15) $3\frac{1}{4}$

16) $3\frac{13}{15}$

17) $3\frac{1}{8}$

18) $2\frac{13}{15}$

19) $3\frac{7}{30}$

20) $4\frac{3}{35}$

21) $5\frac{1}{12}$

22) $2\frac{1}{16}$

23) $4\frac{1}{24}$

24) $2\frac{71}{72}$

25) $5\frac{33}{35}$

26) $6\frac{32}{63}$

Multiplying and Dividing Mixed Numbers

1) $12\frac{3}{8}$

2) $23\frac{1}{9}$

3) $35\frac{15}{16}$

4) $8\frac{2}{3}$

5) 5

6) $6\frac{6}{7}$

7) $30\frac{1}{3}$

8) $7\frac{6}{7}$

9) $21\frac{7}{8}$

10) 11

11) $\frac{4}{7}$

12) $1\frac{1}{4}$

13) $4\frac{2}{9}$

14) $2\frac{23}{24}$

15) 3

16) $\frac{11}{45}$

17) $\frac{19}{84}$

18) $\frac{52}{57}$

19) $4\frac{4}{27}$

20) $2\frac{26}{33}$

21) $3\frac{17}{36}$

22) $\frac{19}{22}$

23) $3\frac{27}{91}$

24) $3\frac{13}{36}$

25) $7\frac{1}{5}$

26) $4\frac{16}{45}$

Chapter 3:
Decimals

Topics that you will practice in this chapter:

- ✓ Adding and Subtracting Decimals
- ✓ Multiplying and Dividing Decimals
- ✓ Comparing Decimals
- ✓ Rounding Decimals

"The study of mathematics, like the Nile, begins in minuteness but ends in magnificence." –
Charles Caleb Colton

ACT Aspire Subject Test Mathematics Grade 7

Adding and Subtracting Decimals

✎ Add and subtract decimals.

1) 35.19 − 24.28 = _____

4) 38.72 − 21.68 = _____

7) 86.09 − 35.14 = _____

2) 34.29 + 42.58 = _____

5) 57.39 + 26.54 = _____

8) 54.51 + 32.66 = _____

3) 61.20 + 33.75 = _____

6) 70.24 − 42.35 = _____

9) 114.21 − 88.69 = _____

✎ Find the missing number.

10) ___ + 2.8 = 5.4

11) 4.1 + ___ = 5.88

12) 6.45 + ___ = 8

13) 7.25 − ___ = 3.40

14) ___ − 2.35 = 4.25

15) ___ − 19.85 = 6.54

16) 22.15 + ___ = 28.95

17) ___ − 37.16 = 9.42

18) ___ + 24.50 = 34.19

19) 72.40 + ___ = 125.20

ACT Aspire Subject Test Mathematics Grade 7

Multiplying and Dividing Decimals

✏ **Find the product.**

1) $0.5 \times 0.6 =$

2) $3.3 \times 0.4 =$

3) $1.28 \times 0.5 =$

4) $0.35 \times 0.6 =$

5) $1.85 \times 0.6 =$

6) $0.24 \times 0.5 =$

7) $5.25 \times 1.4 =$

8) $18.5 \times 4.6 =$

9) $15.4 \times 6.8 =$

10) $19.5 \times 2.6 =$

11) $32.2 \times 1.5 =$

12) $78.4 \times 4.5 =$

✏ **Find the quotient.**

13) $1.85 \div 10 =$

14) $74.6 \div 100 =$

15) $3.6 \div 3 =$

16) $9.6 \div 0.4 =$

17) $15.5 \div 0.5 =$

18) $32.8 \div 0.2 =$

19) $22.15 \div 1{,}000 =$

20) $53.55 \div 0.7 =$

21) $322.2 \div 0.2 =$

22) $50.67 \div 0.18 =$

23) $77.4 \div 0.8 =$

24) $27.93 \div 0.03 =$

ACT Aspire Subject Test Mathematics Grade 7

Comparing Decimals

✎ Write the correct comparison symbol (>, < or =).

1) 0.70 ☐ 0.070

2) 0.049 ☐ 0.49

3) 5.090 ☐ 5.09

4) 2.57 ☐ 2.05

5) 9.03 ☐ 0.930

6) 6.06 ☐ 6.6

7) 7.02 ☐ 7.020

8) 3.04 ☐ 3.2

9) 3.61 ☐ 3.245

10) 0.986 ☐ 0.0986

11) 17.24 ☐ 17.240

12) 0.759 ☐ 0.81

13) 9.040 ☐ 9.40

14) 5.73 ☐ 5.213

15) 9.44 ☐ 9.404

16) 7.17 ☐ 7.170

17) 4.85 ☐ 4.085

18) 9.041 ☐ 9.40

19) 3.033 ☐ 3.030

20) 4.97 ☐ 4.970

WWW.MathNotion.Com

Rounding Decimals

🖎 Round each decimal to the nearest whole number.

1) 28.12   3) 16.22   5) 7.95

2) 6.9   4) 8.5   6) 52.7

🖎 Round each decimal to the nearest tenth.

7) 31.761   9) 94.729   11) 13.219

8) 14.421   10) 77.89   12) 59.89

🖎 Round each decimal to the nearest hundredth.

13) 8.428   15) 55.3786   17) 62.241

14) 23.812   16) 231.912   18) 19.447

🖎 Round each decimal to the nearest thousandth.

19) 15.54324   21) 243.8652   23) 67.1983

20) 34.62586   22) 80.4529   24) 72.36788

ACT Aspire Subject Test Mathematics Grade 7

Answers of Worksheets

Adding and Subtracting Decimals

1) 10.91
2) 76.87
3) 94.95
4) 17.04
5) 83.93
6) 27.89
7) 50.95
8) 87.17
9) 25.52
10) 2.6
11) 1.78
12) 1.55
13) 3.85
14) 6.6
15) 26.39
16) 6.8
17) 46.58
18) 9.69
19) 52.8

Multiplying and Dividing Decimals

1) 0.3
2) 1.32
3) 0.64
4) 0.21
5) 1.11
6) 0.12
7) 7.35
8) 85.1
9) 104.72
10) 50.7
11) 48.3
12) 352.8
13) 0.185
14) 0.746
15) 1.2
16) 24
17) 31
18) 164
19) 0.02215
20) 76.5
21) 1,611
22) 281.5
23) 96.75
24) 931

Comparing Decimals

1) >
2) <
3) =
4) >
5) >
6) <
7) =
8) <
9) >
10) >
11) =
12) <
13) <
14) >
15) >
16) =
17) >
18) <
19) >
20) =

Rounding Decimals

1) 28
2) 7
3) 16
4) 9
5) 8
6) 53
7) 31.8
8) 14.4
9) 94.7
10) 77.9
11) 13.2
12) 59.9
13) 8.43
14) 23.81
15) 55.38
16) 231.91
17) 62.24
18) 19.45
19) 15.543
20) 34.626
21) 243.865
22) 80.453
23) 67.198
24) 72.368

WWW.MathNotion.Com

Chapter 4:
Proportions, Ratios, and Percent

Topics that you will practice in this chapter:

- ✓ Simplifying Ratios
- ✓ Proportional Ratios
- ✓ Similarity and Ratios
- ✓ Ratio and Rates Word Problems
- ✓ Percentage Calculations
- ✓ Percent Problems
- ✓ Discount, Tax and Tip
- ✓ Percent of Change
- ✓ Simple Interest

Without mathematics, there's nothing you can do. Everything around you is mathematics. Everything around you is numbers." – Shakuntala Devi

ACT Aspire Subject Test Mathematics Grade 7

Simplifying Ratios

✎ **Reduce each ratio.**

1) $15:20 = \_\_:\_\_$
2) $7:70 = \_\_:\_\_$
3) $16:28 = \_\_:\_\_$
4) $7:21 = \_\_:\_\_$
5) $4:40 = \_\_:\_\_$
6) $6:48 = \_\_:\_\_$
7) $16:64 = \_\_:\_\_$
8) $10:25 = \_\_:\_\_$

9) $8:48 = \_\_:\_\_$
10) $49:63 = \_\_:\_\_$
11) $18:27 = \_\_:\_\_$
12) $35:10 = \_\_:\_\_$
13) $90:9 = \_\_:\_\_$
14) $24:32 = \_\_:\_\_$
15) $7:56 = \_\_:\_\_$
16) $45:63 = \_\_:\_\_$

17) $56:72 = \_\_:\_\_$
18) $26:13 = \_\_:\_\_$
19) $15:45 = \_\_:\_\_$
20) $28:4 = \_\_:\_\_$
21) $24:48 = \_\_:\_\_$
22) $30:24 = \_\_:\_\_$
23) $70:140 = \_\_:\_\_$
24) $6:180 = \_\_:\_\_$

✎ **Write each ratio as a fraction in simplest form.**

25) $6:12 =$
26) $30:50 =$
27) $15:35 =$
28) $9:27 =$
29) $8:24 =$
30) $18:84 =$
31) $7:14 =$

32) $7:35 =$
33) $40:96 =$
34) $12:54 =$
35) $44:52 =$
36) $12:27 =$
37) $15:180 =$
38) $39:143 =$

39) $20:300 =$
40) $30:120 =$
41) $56:42 =$
42) $26:130 =$
43) $66:123 =$
44) $70:630 =$
45) $75:125 =$

ACT Aspire Subject Test Mathematics Grade 7

Proportional Ratios

✎ Fill in the blanks; Calculate each proportion.

1) $3:8 = \_\_ : 48$
2) $2:5 = 20:\_\_$
3) $1:9 = \_\_ : 81$
4) $6:7 = 12:\_\_$
5) $9:2 = 63:\_\_$
6) $8:7 = \_\_ : 49$
7) $20:3 = \_\_ : 15$
8) $1:3 = \_\_ : 75$
9) $7:6 = \_\_ : 60$
10) $8:5 = \_\_ : 45$
11) $3:10 = 60:\_\_$
12) $6:11 = 42:\_\_$

✎ State if each pair of ratios form a proportion.

13) $\frac{3}{20}$ and $\frac{9}{60}$
14) $\frac{1}{7}$ and $\frac{6}{42}$
15) $\frac{3}{7}$ and $\frac{24}{56}$
16) $\frac{4}{9}$ and $\frac{12}{18}$
17) $\frac{1}{9}$ and $\frac{12}{81}$
18) $\frac{7}{8}$ and $\frac{21}{28}$
19) $\frac{9}{13}$ and $\frac{27}{39}$
20) $\frac{1}{8}$ and $\frac{8}{64}$
21) $\frac{6}{19}$ and $\frac{30}{85}$
22) $\frac{5}{9}$ and $\frac{40}{81}$
23) $\frac{9}{14}$ and $\frac{108}{168}$
24) $\frac{15}{23}$ and $\frac{360}{552}$

✎ Calculate each proportion.

25) $\frac{20}{25} = \frac{32}{x}, x = \_\_\_$
26) $\frac{1}{8} = \frac{32}{x}, x = \_\_\_$
27) $\frac{15}{5} = \frac{21}{x}, x = \_\_\_$
28) $\frac{1}{7} = \frac{x}{294}, x = \_\_\_$
29) $\frac{7}{9} = \frac{x}{81}, x = \_\_\_$
30) $\frac{1}{5} = \frac{13}{x}, x = \_\_\_$
31) $\frac{9}{5} = \frac{36}{x}, x = \_\_\_$
32) $\frac{6}{13} = \frac{48}{x}, x = \_\_\_$
33) $\frac{5}{8} = \frac{x}{88}, x = \_\_\_$
34) $\frac{4}{15} = \frac{x}{240}, x = \_\_\_$
35) $\frac{9}{19} = \frac{x}{266}, x = \_\_\_$
36) $\frac{7}{15} = \frac{x}{270}, x = \_\_\_$

ACT Aspire Subject Test Mathematics Grade 7

Similarity and Ratios

✎ **Each pair of figures is similar. Find the missing side.**

1) 18, ?, 12 and 6, 5, 4

2) 12, 9, 7.5 and 8, 6, ?

3) 44, 60 and 11, ?

4) ?, 7 and 104, 56

✎ **Calculate.**

5) Two rectangles are similar. The first is 24 feet wide and 120 feet long. The second is 30 feet wide. What is the length of the second rectangle? _____

6) Two rectangles are similar. One is 5 meters by 36 meters. The longer side of the second rectangle is 90 meters. What is the other side of the second rectangle? _____

7) A building casts a shadow 25 ft long. At the same time a girl 10 ft tall casts a shadow 5 ft long. How tall is the building? _____

8) The scale of a map of Texas is 4 inches: 32 miles. If you measure the distance from Dallas to Martin County as 38.4 inches, approximately how far is Martin County from Dallas? _____

ACT Aspire Subject Test Mathematics Grade 7

Ratio and Rates Word Problems

✏ **Find the answer for each word problem.**

1) Mason has 24 red cards and 36 green cards. What is the ratio of Mason's red cards to his green cards? _____

2) In a party, 45 soft drinks are required for every 54 guests. If there are 378 guests, how many soft drinks is required? _____

3) In Mason's class, 42 of the students are tall and 24 are short. In Michael's class 84 students are tall and 48 students are short. Which class has a higher ratio of tall to short students? _____

4) The price of 5 apples at the Quick Market is $4.6. The price of 7 of the same apples at Walmart is $5.95. Which place is the better buy? _____

5) The bakers at a Bakery can make 90 bagels in 3 hours. How many bagels can they bake in 24 hours? What is that rate per hour? _____

6) You can buy 5 cans of green beans at a supermarket for $5.75. How much does it cost to buy 45 cans of green beans? _____

7) The ratio of boys to girls in a class is 4: 7. If there are 32 boys in the class, how many girls are in that class? _____

8) The ratio of red marbles to blue marbles in a bag is 3: 7. If there are 50 marbles in the bag, how many of the marbles are red? _____

WWW.MathNotion.Com

ACT Aspire Subject Test Mathematics Grade 7

Percentage Calculations

✎ Calculate the given percent of each value.

1) 3% of 60 = ____
2) 20% of 32 = ____
3) 4% of 72 = ____
4) 16% of 32 = ____
5) 25% of 124 = ____
6) 35% of 56 = ____

7) 15% of 20 = ____
8) 14% of 150 = ____
9) 80% of 50 = ____
10) 12% of 115 = ____
11) 72% of 250 = ____
12) 52% of 500 = ____

13) 70% of 400 = ____
14) 27% of 145 = ____
15) 90% of 64 = ____
16) 60% of 55 = ____
17) 22% of 210 = ____
18) 8% of 235 = ____

✎ Calculate the percent of each given value.

19) ___% of 25 = 5
20) ___% of 40 = 20
21) ___% of 25 = 2
22) ___% of 50 = 16
23) ___% of 250 = 5

24) ___% of 40 = 32
25) ___% of 125 = 20
26) ___% of 700 = 49
27) ___% of 350 = 49
28) ___% of 500 = 210

✎ Calculate each percent problem.

29) A Cinema has 250 seats. 60 seats were sold for the current movie. What percent of seats are empty? ____ %

30) There are 68 boys and 92 girls in a class. 75% of the students in the class take the bus to school. How many students do not take the bus to school? ____

WWW.MathNotion.Com

ACT Aspire Subject Test Mathematics Grade 7

Percent Problems

✎ **Calculate each problem.**

1) 9 is what percent of 45? ____%

2) 60 is what percent of 120? ____%

3) 10 is what percent of 200? ____%

4) 15 is what percent of 125? ____%

5) 10 is what percent of 400? ____%

6) 66 is what percent of 55? ____%

7) 40 is what percent of 160? ____%

8) 40 is what percent of 50? ____%

9) 120 is what percent of 800? ____%

10) 78 is what percent of 120? ___%

11) 36 is what percent of 144? ___%

12) 17 is what percent of 85? ___%

13) 90 is what percent of 900? ___%

14) 36 is what percent of 16? ___%

15) 63 is what percent of 14? ___%

16) 18 is what percent of 60? ___%

17) 126 is what percent of 200? ___%

18) 232 is what percent of 40? ___%

✎ **Calculate each percent word problem.**

19) There are 40 employees in a company. On a certain day, 25 were present. What percent showed up for work? ____%

20) A metal bar weighs 60 ounces. 25% of the bar is gold. How many ounces of gold are in the bar? _____

21) A crew is made up of 12 women; the rest are men. If 15% of the crew are women, how many people are in the crew? _____

22) There are 40 students in a class and 8 of them are girls. What percent are boys? ____%

23) The Royals softball team played 400 games and won 280 of them. What percent of the games did they lose? ____%

ACT Aspire Subject Test Mathematics Grade 7

Discount, Tax and Tip

✏️ **Find the selling price of each item.**

1) Original price of a computer: $420
   Tax: 8% Selling price: $_____

2) Original price of a laptop: $280
   Tax: 4% Selling price: $_____

3) Original price of a sofa: $820
   Tax: 5% Selling price: $_____

4) Original price of a car: $15,800
   Tax: 3.6%   Selling price: $_____

5) Original price of a Table: $250
   Tax: 9% Selling price: $_____

6) Original price of a house: $630,000
   Tax: 1.8%   Selling price: $_____

7) Original price of a tablet: $450
   Discount: 30%   Selling price: $____

8) Original price of a chair: $390
   Discount: 8%   Selling price: $____

9) Original price of a book: $75
   Discount: 42%   Selling price: $____

10) Original price of a cellphone: $820
    Discount: 23%   Selling price: $___

11) Food bill: $45
    Tip: 15%   Price: $_____

12) Food bill: $32
    Tipp: 20%   Price: $_____

13) Food bill: $90
    Tip: 35%   Price: $_____

14) Food bill: $42
    Tipp: 12%   Price: $_____

✏️ **Find the answer for each word problem.**

15) Nicolas hired a moving company. The company charged $500 for its services, and Nicolas gives the movers a 40% tip. How much does Nicolas tip the movers? $_____

16) Mason has lunch at a restaurant and the cost of his meal is $90. Mason wants to leave a 25% tip. What is Mason's total bill including tip? $_____

17) The sales tax in Texas is 19.80% and an item costs $350. How much is the tax? $_____

18) The price of a table at Best Buy is $680. If the sales tax is 5%, what is the final price of the table including tax? $_____

WWW.MathNotion.Com

ACT Aspire Subject Test Mathematics Grade 7

Percent of Change

✎ **Find each percent of change.**

1) From 150 to 450. ___ %

2) From 50 ft to 250 ft. ___ %

3) From $60 to $360. ___ %

4) From 60 cm to 180 cm. ___ %

5) From 15 to 45. ___ %

6) From 80 to 16. ___ %

7) From 120 to 360. ___ %

8) From 900 to 450. ___ %

9) From 1,000 to 200. ___ %

10) From 144 to 36. ___ %

✎ **Calculate each percent of change word problem.**

11) Bob got a raise, and his hourly wage increased from $42 to $63. What is the percent increase? ____ %

12) The price of a pair of shoes increases from $50 to $61. What is the percent increase? ___ %

13) At a coffee shop, the price of a cup of coffee increased from $4.80 to $5.76. What is the percent increase in the cost of the coffee? ____ %

14) 51 cm are cut from 85 cm board. What is the percent decrease in length? ____ %

15) In a class, the number of students has been increased from 54 to 81. What is the percent increase? ____ %

16) The price of gasoline rises from $24.40 to $30.50 in one month. By what percent did the gas price rise? ____ %

17) A shirt was originally priced at $38. It went on sale for $24.70. What was the percent that the shirt was discounted? ____ %

WWW.MathNotion.Com

ACT Aspire Subject Test Mathematics Grade 7

Simple Interest

✎ **Determine the simple interest for these loans.**

1) $480 at 11% for 3 years. $ _____

2) $4,200 at 7% for 4 years. $ _____

3) $2,500 at 20% for 3 years. $ _____

4) $6,800 at 3.9% for 4 months. $ _____

5) $800 at 6% for 7 months. $ _____

6) $36,000 at 4.2% for 6 years. $ _____

7) $6,500 at 7% for 4 years. $ _____

8) $850 at 9.5% for 2 years. $ _____

9) $1,200 at 5.8% for 9 months. $ _____

10) $3,000 at 4.5% for 7 years. $ _____

✎ **Calculate each simple interest word problem.**

11) A new car, valued at $22,000, depreciates at 8.5% per year. What is the value of the car one year after purchase? $_____

12) Sara puts $9,000 into an investment yielding 6% annual simple interest; she left the money in for three years. How much interest does Sara get at the end of those three years? $_____

13) A bank is offering 12% simple interest on a savings account. If you deposit $16,400, how much interest will you earn in two years? $_____

14) $720 interest is earned on a principal of $6,000 at a simple interest rate of 4% interest per year. For how many years was the principal invested? _____

15) In how many years will $2,200 yield an interest of $440 at 4% simple interest? _____

16) Jim invested $8,000 in a bond at a yearly rate of 4.5%. He earned $1,440 in interest. How long was the money invested? _____

ACT Aspire Subject Test Mathematics Grade 7

Answers of Worksheets

Simplifying Ratios

1) 3 : 4
2) 1 : 10
3) 4 : 7
4) 1 : 3
5) 1 : 10
6) 1 : 8
7) 2 : 8
8) 2 : 5
9) 1 : 6
10) 7 : 9
11) 2 : 3
12) 7 : 2
13) 10 : 1
14) 3 : 4
15) 1 : 8
16) 5 : 7
17) 7 : 9
18) 2 : 1
19) 1 : 3
20) 7 : 1
21) 1 : 2
22) 5 : 4
23) 1 : 2
24) 1 : 30
25) $\frac{1}{2}$
26) $\frac{3}{5}$
27) $\frac{3}{7}$
28) $\frac{1}{3}$
29) $\frac{1}{3}$
30) $\frac{3}{14}$
31) $\frac{1}{2}$
32) $\frac{1}{5}$
33) $\frac{5}{12}$
34) $\frac{2}{9}$
35) $\frac{11}{13}$
36) $\frac{4}{9}$
37) $\frac{1}{12}$
38) $\frac{3}{11}$
39) $\frac{1}{15}$
40) $\frac{1}{4}$
41) $\frac{4}{3}$
42) $\frac{1}{5}$
43) $\frac{22}{41}$
44) $\frac{1}{9}$
45) $\frac{3}{5}$

Proportional Ratios

1) 18
2) 50
3) 9
4) 14
5) 14
6) 56
7) 100
8) 25
9) 70
10) 72
11) 200
12) 77
13) Yes
14) Yes
15) Yes
16) No
17) No
18) No
19) Yes
20) Yes
21) No
22) No
23) Yes
24) Yes
25) 40
26) 256
27) 7
28) 42
29) 63
30) 65
31) 20
32) 104
33) 55
34) 64
35) 126
36) 126

Similarity and ratios

1) 15
2) 5
3) 15
4) 13
5) 150 feet
6) 12.5 meters
7) 50 feet
8) 307.2 miles

Ratio and Rates Word Problems

1) 2 : 3
2) 315

WWW.MathNotion.Com

ACT Aspire Subject Test Mathematics Grade 7

3) The ratio for both classes is 7 to 4.
4) Walmart is a better buy.
5) 720, the rate is 30 per hour.
6) $51.75
7) 56
8) 15

Percentage Calculations

1) 1.8
2) 6.4
3) 2.88
4) 5.12
5) 31
6) 19.6
7) 3
8) 21
9) 40
10) 13.8
11) 180
12) 260
13) 280
14) 39.15
15) 57.6
16) 33
17) 46.2
18) 18.8
19) 20%
20) 50%
21) 8%
22) 32%
23) 2%
24) 80%
25) 16%
26) 7%
27) 14%
28) 42%
29) 76%
30) 40

Percent Problems

1) 20%
2) 50%
3) 5%
4) 12%
5) 2.5%
6) 120%
7) 25%
8) 80%
9) 15%
10) 65%
11) 25%
12) 20%
13) 10%
14) 225%
15) 450%
16) 30%
17) 63%
18) 580%
19) 62.5%
20) 15 ounces
21) 80
22) 80%
23) 30%

Discount, Tax and Tip

1) $453.60
2) $291.20
3) $861.00
4) $16,368.80
5) $272.50
6) $641,340
7) $315.00
8) $358.80
9) $43.50
10) $631.40
11) $51.75
12) $38.40
13) $121.50
14) $47.04
15) $200.00
16) $112.50
17) $69.30
18) $714.00

ACT Aspire Subject Test Mathematics Grade 7

Percent of Change

1) 200%
2) 400%
3) 500%
4) 200%
5) 200%
6) 80%
7) 200%
8) 50%
9) 80%
10) 75%
11) 50%
12) 22%
13) 20%
14) 60%
15) 50%
16) 25%
17) 35%

Simple Interest

1) $158.40
2) $1,176.00
3) $1,500.00
4) $88.40
5) $28.00
6) $9,072.00
7) $1,820.00
8) $161.50
9) $52.20
10) $945.00
11) $20,130.00
12) $1,620.00
13) $3,936.00
14) 3 years
15) 5 years
16) 4 years

Chapter 5:

Exponents and Radicals Expressions

Topics that you will practice in this chapter:

- ✓ Multiplication Property of Exponents
- ✓ Zero and Negative Exponents
- ✓ Division Property of Exponents
- ✓ Powers of Products and Quotients
- ✓ Negative Exponents and Negative Bases
- ✓ Scientific Notation
- ✓ Square Roots

Mathematics is no more computation than typing is literature.

– John Allen Paulos

Multiplication Property of Exponents

✏️ **Simplify and write the answer in exponential form.**

1) $4 \times 4^5 =$

2) $8^4 \times 8 =$

3) $7^3 \times 7^3 =$

4) $9^2 \times 9^2 =$

5) $2^2 \times 2^4 \times 2 =$

6) $5 \times 5^3 \times 5^3 =$

7) $4^3 \times 4^2 \times 4 \times 4 =$

8) $5x \times x =$

9) $x^3 \times x^3 =$

10) $x^7 \times x^2 =$

11) $x^4 \times x^3 \times x^2 =$

12) $10x \times 3x =$

13) $4x^3 \times 4x^3 =$

14) $7x^3 \times x =$

15) $3x^2 \times 4x^2 \times x^2 =$

16) $5x^4 \times x^4 =$

17) $2x^8 \times 2x =$

18) $6x \times x^5 =$

19) $4x^2 \times 6x^6 =$

20) $5yx^3 \times 4x =$

21) $7x^3 \times y^5 x^7 =$

22) $y^2 x^3 \times y^5 x^4 =$

23) $3x^5 \times 4x^3 y^4 =$

24) $4x^4 \times 9x^2 y^5 =$

25) $5x^3 y^4 \times 6x^8 y^2 =$

26) $8x^3 y^6 \times 4xy^3 =$

27) $2xy^5 \times 6x^3 y^3 =$

28) $4x^5 y^2 \times 4x^2 y^8 =$

29) $7x \times 3y^8 x^2 \times y^5 =$

30) $x^3 \times 2y^3 x^4 \times 2y =$

31) $3yx^4 \times 3y^4 x \times 3xy^3 =$

32) $6y^3 \times 2y^2 x^4 \times 10yx^5 =$

ACT Aspire Subject Test Mathematics Grade 7

Zero and Negative Exponents

✏ **Evaluate the following expressions.**

1) $1^{-5} =$

2) $4^{-1} =$

3) $0^{10} =$

4) $1^{15} =$

5) $5^{-2} =$

6) $3^{-3} =$

7) $9^{-1} =$

8) $10^{-2} =$

9) $12^{-2} =$

10) $2^{-5} =$

11) $3^{-4} =$

12) $2^{-4} =$

13) $6^{-3} =$

14) $10^{-3} =$

15) $30^{-1} =$

16) $15^{-2} =$

17) $4^{-3} =$

18) $2^{-7} =$

19) $5^{-3} =$

20) $4^{-4} =$

21) $3^{-5} =$

22) $10^{-4} =$

23) $2^{-10} =$

24) $8^{-3} =$

25) $20^{-2} =$

26) $14^{-2} =$

27) $9^{-3} =$

28) $100^{-2} =$

29) $5^{-4} =$

30) $4^{-6} =$

31) $\left(\frac{1}{4}\right)^{-3} =$

32) $\left(\frac{1}{6}\right)^{-2} =$

33) $\left(\frac{1}{7}\right)^{-2} =$

34) $\left(\frac{2}{3}\right)^{-3} =$

35) $\left(\frac{1}{13}\right)^{-2} =$

36) $\left(\frac{7}{12}\right)^{-2} =$

37) $\left(\frac{1}{6}\right)^{-3} =$

38) $\left(\frac{1}{300}\right)^{-2} =$

39) $\left(\frac{2}{9}\right)^{-2} =$

40) $\left(\frac{7}{5}\right)^{-1} =$

41) $\left(\frac{13}{23}\right)^{0} =$

42) $\left(\frac{1}{4}\right)^{-5} =$

WWW.MathNotion.Com

ACT Aspire Subject Test Mathematics Grade 7

Division Property of Exponents

✎ **Simplify.**

1) $\dfrac{5^6}{5^7} =$

2) $\dfrac{8^8}{8^6} =$

3) $\dfrac{4^5}{4} =$

4) $\dfrac{3}{3^5} =$

5) $\dfrac{x}{x^6} =$

6) $\dfrac{3 \times 3^2}{3^2 \times 3^5} =$

7) $\dfrac{9^4}{9^2} =$

8) $\dfrac{10 \times 10^9}{10^2 \times 10^7} =$

9) $\dfrac{7^5 \times 7^7}{7^4 \times 7^8} =$

10) $\dfrac{15x}{30x^6} =$

11) $\dfrac{3x^9}{4x^4} =$

12) $\dfrac{15x^8}{10x^9} =$

13) $\dfrac{42x^5}{6y^9} =$

14) $\dfrac{36y^8}{4x^4 y^5} =$

15) $\dfrac{2x^7}{9x} =$

16) $\dfrac{49x^8 y^6}{7x^9} =$

17) $\dfrac{48x^2}{24x^6 y^{12}} =$

18) $\dfrac{30yx^5}{6yx^7} =$

19) $\dfrac{19x^7 y}{38x^{12} y^4} =$

20) $\dfrac{9x^8}{63x^8} =$

21) $\dfrac{9x^{-9}}{4x^{-3}} =$

ACT Aspire Subject Test Mathematics Grade 7

Powers of Products and Quotients

✏️ **Simplify.**

1) $(4^3)^2 =$

2) $(2^3)^4 =$

3) $(2 \times 2^3)^2 =$

4) $(5 \times 5^5)^6 =$

5) $(19^4 \times 19^2)^3 =$

6) $(2^3 \times 2^4)^4 =$

7) $(5 \times 5^2)^2 =$

8) $(4^4)^4 =$

9) $(8x^5)^2 =$

10) $(3x^2 y^4)^4 =$

11) $(7x^5 y^2)^2 =$

12) $(5x^4 y^4)^3 =$

13) $(2x^3 y^3)^5 =$

14) $(10x^3 y^4)^3 =$

15) $(13y^3 y)^2 =$

16) $(5x^6 x^4)^2 =$

17) $(6x^7 y^6)^3 =$

18) $(12x^5 x^7)^2 =$

19) $(2x^4 \times 2x)^4 =$

20) $(2x^4 y^3)^5 =$

21) $(15x^7 y^2)^2 =$

22) $(8x^3 y^5)^3 =$

23) $(3x \times 2y^2)^4 =$

24) $\left(\frac{4x}{x^5}\right)^2 =$

25) $\left(\frac{x^4 y^5}{x^3 y^5}\right)^9 =$

26) $\left(\frac{36xy}{6x^5}\right)^3 =$

27) $\left(\frac{x^7}{x^8 y^2}\right)^6 =$

28) $\left(\frac{xy^4}{x^3 y^6}\right)^{-3} =$

29) $\left(\frac{5xy^8}{x^3}\right)^2 =$

30) $\left(\frac{xy^6}{2xy^3}\right)^{-4} =$

ACT Aspire Subject Test Mathematics Grade 7

Negative Exponents and Negative Bases

✏️ **Simplify.**

1) $-9^{-1} =$

2) $-9^{-2} =$

3) $-2^{-5} =$

4) $-x^{-7} =$

5) $11x^{-1} =$

6) $-8x^{-3} =$

7) $-12x^{-5} =$

8) $-9x^{-8}y^{-6} =$

9) $32x^{-5}y^{-1} =$

10) $10a^{-9}b^{-3} =$

11) $-17x^4y^{-6} =$

12) $-\dfrac{25}{x^{-5}} =$

13) $-\dfrac{13x}{a^{-7}} =$

14) $\left(-\dfrac{1}{3}\right)^{-4} =$

15) $\left(-\dfrac{3}{4}\right)^{-2} =$

16) $-\dfrac{14}{a^{-6}b^{-3}} =$

17) $-\dfrac{7x}{x^{-8}} =$

18) $-\dfrac{a^{-9}}{b^{-5}} =$

19) $-\dfrac{11}{x^{-5}} =$

20) $\dfrac{8b}{-16c^{-6}} =$

21) $\dfrac{12ab}{a^{-4}b^{-3}} =$

22) $-\dfrac{8n^{-4}}{32p^{-7}} =$

23) $\dfrac{16ab^{-6}}{-6c^{-5}} =$

24) $\left(\dfrac{10a}{5c}\right)^{-4} =$

25) $\left(-\dfrac{12x}{4yz}\right)^{-3} =$

26) $\dfrac{8ab^{-7}}{-5c^{-3}} =$

27) $\left(-\dfrac{x^4}{x^5}\right)^{-5} =$

28) $\left(-\dfrac{x^{-2}}{7x^3}\right)^{-2} =$

29) $\left(-\dfrac{x^{-4}}{x^2}\right)^{-6} =$

WWW.MathNotion.Com

ACT Aspire Subject Test Mathematics Grade 7

Scientific Notation

✎ Write each number in scientific notation.

1) $0.223 =$

2) $0.09 =$

3) $4.5 =$

4) $900 =$

5) $2,000 =$

6) $0.006 =$

7) $33 =$

8) $9,400 =$

9) $1,470 =$

10) $52,000 =$

11) $8,000,000 =$

12) $0.00009 =$

13) $2,158,000 =$

14) $0.0039 =$

15) $0.000075 =$

16) $4,300,000 =$

17) $130,000 =$

18) $4,000,000,000 =$

19) $0.00009 =$

20) $0.0039 =$

✎ Write each number in standard notation.

21) $4 \times 10^{-1} =$

22) $1.2 \times 10^{-3} =$

23) $2.7 \times 10^5 =$

24) $6 \times 10^{-4} =$

25) $3.6 \times 10^{-3} =$

26) $5.5 \times 10^5 =$

27) $3.2 \times 10^4 =$

28) $3.88 \times 10^6 =$

29) $7 \times 10^{-6} =$

30) $4.2 \times 10^{-7} =$

WWW.MathNotion.Com

ACT Aspire Subject Test Mathematics Grade 7

Square Roots

✎ **Find the value each square root.**

1) $\sqrt{16} =$ ___

2) $\sqrt{25} =$ ___

3) $\sqrt{1} =$ ___

4) $\sqrt{64} =$ ___

5) $\sqrt{0} =$ ___

6) $\sqrt{196} =$ ___

7) $\sqrt{4} =$ ___

8) $\sqrt{256} =$ ___

9) $\sqrt{36} =$ ___

10) $\sqrt{289} =$ ___

11) $\sqrt{169} =$ ___

12) $\sqrt{144} =$ ___

13) $\sqrt{100} =$ ___

14) $\sqrt{1,600} =$ ___

15) $\sqrt{2,500} =$ ___

16) $\sqrt{324} =$ ___

17) $\sqrt{529} =$ ___

18) $\sqrt{20} =$ ___

19) $\sqrt{625} =$ ___

20) $\sqrt{18} =$ ___

21) $\sqrt{50} =$ ___

22) $\sqrt{1,024} =$ ___

23) $\sqrt{160} =$ ___

24) $\sqrt{32} =$ ___

✎ **Evaluate.**

25) $\sqrt{4} \times \sqrt{25} =$ _____

26) $\sqrt{36} \times \sqrt{49} =$ _____

27) $\sqrt{6} \times \sqrt{6} =$ _____

28) $\sqrt{13} \times \sqrt{13} =$ _____

29) $2\sqrt{5} \times 3\sqrt{5} =$ _____

30) $\sqrt{12} \times \sqrt{3} =$ _____

31) $\sqrt{13} + \sqrt{13} =$ _____

32) $\sqrt{10} + 2\sqrt{10} =$ _____

33) $12\sqrt{7} - 10\sqrt{7} =$ _____

34) $4\sqrt{10} \times 2\sqrt{10} =$ _____

35) $5\sqrt{3} \times 8\sqrt{3} =$ _____

36) $6\sqrt{3} - \sqrt{12} =$ _____

ACT Aspire Subject Test Mathematics Grade 7

Answers of Worksheets

Multiplication Property of Exponents

1) 4^6
2) 8^5
3) 7^6
4) 9^4
5) 2^7
6) 5^7
7) 4^7
8) $5x^2$
9) x^6
10) x^9
11) x^9
12) $30x^2$
13) $16x^6$
14) $7x^4$
15) $12x^6$
16) $5x^8$
17) $4x^9$
18) $6x^6$
19) $24x^8$
20) $20x^4y$
21) $7x^{10}y^5$
22) x^7y^7
23) $12x^8y^4$
24) $36x^6y^5$
25) $30x^{11}y^6$
26) $32x^4y^9$
27) $12x^4y^8$
28) $16x^7y^{10}$
29) $21x^3y^{13}$
30) $4x^7y^4$
31) $27x^6y^8$
32) $120x^9y^6$

Zero and Negative Exponents

1) 1
2) $\frac{1}{4}$
3) 0
4) 1
5) $\frac{1}{25}$
6) $\frac{1}{27}$
7) $\frac{1}{9}$
8) $\frac{1}{100}$
9) $\frac{1}{144}$
10) $\frac{1}{32}$
11) $\frac{1}{81}$
12) $\frac{1}{16}$
13) $\frac{1}{216}$
14) $\frac{1}{1,000}$
15) $\frac{1}{30}$
16) $\frac{1}{225}$
17) $\frac{1}{64}$
18) $\frac{1}{128}$
19) $\frac{1}{125}$
20) $\frac{1}{256}$
21) $\frac{1}{243}$
22) $\frac{1}{10,000}$
23) $\frac{1}{1,024}$
24) $\frac{1}{512}$
25) $\frac{1}{400}$
26) $\frac{1}{196}$
27) $\frac{1}{729}$
28) $\frac{1}{10,000}$
29) $\frac{1}{625}$
30) $\frac{1}{4,096}$
31) 64
32) 36
33) 49
34) $\frac{27}{8}$
35) 169
36) $\frac{144}{49}$
37) 216
38) $90,000$
39) $\frac{81}{4}$
40) $\frac{5}{7}$
41) 1
42) $1,024$

Division Property of Exponents

1) $\frac{1}{5}$
2) 8^2
3) 4^4
4) $\frac{1}{3^4}$
5) $\frac{1}{x^5}$
6) $\frac{1}{3^4}$
7) 9^2
8) 10
9) 1
10) $\frac{1}{2x^5}$
11) $\frac{3x^5}{4}$
12) $\frac{3}{2x}$
13) $\frac{7x^5}{y^9}$
14) $\frac{9y^3}{x^4}$
15) $\frac{2x^6}{9}$

ACT Aspire Subject Test Mathematics Grade 7

16) $\frac{7y^6}{x}$ 18) $\frac{5}{x^2}$ 19) $\frac{1}{2x^5y^3}$ 21) $\frac{9}{4x^6}$

17) $\frac{2}{x^4y^{12}}$ 20) $\frac{1}{7}$

Powers of Products and Quotients

1) 4^6
2) 2^{12}
3) 2^8
4) 5^{36}
5) 19^{18}
6) 2^{28}
7) 5^6
8) 4^{16}
9) $64x^{10}$
10) $81x^8y^{16}$
11) $49x^{10}y^4$
12) $125x^{12}y^{12}$
13) $32x^{15}y^{15}$
14) $1{,}000x^9y^{12}$
15) $169y^8$
16) $25x^{20}$
17) $216x^{21}y^{18}$
18) $144x^{24}$
19) $256x^{20}$
20) $32x^{20}y^{15}$
21) $225x^{14}y^4$
22) $512x^9y^{15}$
23) $1{,}296x^4y^8$
24) $\frac{16}{x^8}$
25) x^9
26) $\frac{216y^3}{x^{12}}$
27) $\frac{1}{x^6y^{12}}$
28) x^6y^6
29) $\frac{25y^{16}}{x^4}$
30) $\frac{16}{y^{12}}$

Negative Exponents and Negative Bases

1) $-\frac{1}{9}$
2) $-\frac{1}{81}$
3) $-\frac{1}{32}$
4) $-\frac{1}{x^7}$
5) $\frac{11}{x}$
6) $-\frac{8}{x^3}$
7) $-\frac{12}{x^5}$
8) $-\frac{9}{x^8y^6}$
9) $\frac{32}{x^5y}$
10) $\frac{10}{a^9b^3}$
11) $-\frac{17x^4}{y^6}$
12) $-25x^5$
13) $-13xa^7$
14) 81
15) $\frac{16}{9}$
16) $-14a^6b^3$
17) $-7x^9$
18) $-\frac{b^5}{a^9}$
19) $-11x^5$
20) $-\frac{bc^6}{2}$
21) $12a^5b^4$
22) $-\frac{p^7}{4n^4}$
23) $-\frac{8ac^5}{3b^6}$
24) $\frac{c^4}{16a^4}$
25) $\frac{y^3z^3}{27x^3}$
26) $-\frac{8ac^3}{5b^7}$
27) $-x^5$
28) $49x^{10}$
29) x^{36}

Scientific Notation

ACT Aspire Subject Test Mathematics Grade 7

1) 2.23×10^{-1}
2) 9×10^{-2}
3) 4.5×10^{0}
4) 9×10^{2}
5) 2×10^{3}
6) 6×10^{-3}
7) 3.3×10^{1}
8) 9.4×10^{3}
9) 1.47×10^{3}
10) 5.2×10^{4}
11) 8×10^{6}
12) 9×10^{-5}
13) 2.158×10^{6}
14) 3.9×10^{-3}
15) 7.5×10^{-5}
16) 4.3×10^{6}
17) 1.3×10^{5}
18) 4×10^{9}
19) 9×10^{-5}
20) 3.9×10^{-3}
21) 0.4
22) 0.0012
23) 270,000
24) 0.0006
25) 0.0036
26) 550,000
27) 32,000
28) 3,880,000
29) 0.000007
30) 0.00000042

Square Roots

1) 4
2) 5
3) 1
4) 8
5) 0
6) 14
7) 2
8) 16
9) 6
10) 17
11) 13
12) 12
13) 10
14) 40
15) 50
16) 18
17) 23
18) $2\sqrt{5}$
19) 25
20) $3\sqrt{2}$
21) $5\sqrt{2}$
22) 32
23) $4\sqrt{10}$
24) $4\sqrt{2}$
25) 10
26) 42
27) 6
28) 13
29) 30
30) 6
31) $2\sqrt{13}$
32) $3\sqrt{10}$
33) $2\sqrt{7}$
34) 80
35) 120
36) $4\sqrt{3}$

Chapter 6:
Algebraic Expressions

Topics that you will practice in this chapter:

- ✓ Translate Phrases into an Algebraic Statement
- ✓ Simplifying Variable Expressions
- ✓ The Distributive Property
- ✓ Evaluating One Variable Expressions
- ✓ Evaluating Two Variables Expressions
- ✓ Combining like Terms

Mathematics is, as it were, a sensuous logic, and relates to philosophy as do the arts, music, and plastic art to poetry. — *K. Shegel*

ACT Aspire Subject Test Mathematics Grade 7

Translate Phrases into an Algebraic Statement

✎ Write an algebraic expression for each phrase.

1) 9 multiplied by x. _____

2) Subtract 11 from y. _____

3) 19 divided by x. _____

4) 38 decreased by y. _____

5) Add y to 40. _____

6) The square of 6. _____

7) x raised to the fifth power. _____

8) The sum of six and a number. _____

9) The difference between fifty-seven and y. _____

10) The quotient of nine and a number. _____

11) The quotient of the square of x and 25. _____

12) The difference between x and 6 is 19. _____

13) 10 times a reduced by the square of b. _____

14) Subtract the product of a and b from 41. _____

ACT Aspire Subject Test Mathematics Grade 7

Simplifying Variable Expressions

✎ **Simplify each expression.**

1) $3(x + 5) =$

2) $(-4)(7x - 5) =$

3) $11x + 5 - 6x =$

4) $-4 - 2x^2 - 6x^2 =$

5) $7 + 13x^2 + 3 =$

6) $3x^2 + 7x + 15x^2 =$

7) $3x^2 - 12x^2 + 4x =$

8) $4x^2 - 8x - 2x =$

9) $6x + 7(3 - 4x) =$

10) $8x + 4(15x - 3) =$

11) $6(-3x - 9) - 17 =$

12) $-11x^2 - (-5x) =$

13) $2x + 7 + 5 - 8x =$

14) $7 + 6x - 11 - 5x =$

15) $27x + 8 - 13 - 5x =$

16) $(-11)(-5x + 2) - 41x =$

17) $19x - 4(4 - 2x) =$

18) $16x + 3(3x + 6) + 10 =$

19) $5(-2x - 4) - 13x =$

20) $16x - 3x(x + 10) =$

21) $17x + 5x(2 - 4x) =$

22) $5x(-4x - 7) + 20x =$

23) $25x - 19 + 4x^2 =$

24) $6x(x - 11) + 25 =$

25) $4x - 5 + 15x + 3x^2 =$

26) $-7x^2 - 11x - 9x =$

27) $10x - 9x^2 - 3x^2 - 7 =$

28) $13 + 3x^2 - 9x^2 - 21x =$

29) $22x + 10x^2 - 15x + 17 =$

30) $4x^2 + 25x + 21x^2 =$

31) $29 - 12x^2 - 23x - 4x^2 =$

32) $22x - 19x - 9x^2 + 30 =$

WWW.MathNotion.Com

The Distributive Property

✋ Use the distributive property to simply each expression.

1) $4(1 + 2x) =$

2) $2(4 + 7x) =$

3) $3(4x - 4) =$

4) $(2x - 5)(-6) =$

5) $(-3)(x + 6) =$

6) $(4 + 3x)2 =$

7) $(-5)(8 - 3x) =$

8) $-(-5 - 7x) =$

9) $(-6x + 3)(-3) =$

10) $(-4)(x - 7) =$

11) $-(5 - 3x) =$

12) $3(9 + 4x) =$

13) $6(4 + 3x) =$

14) $(-5x + 3)2 =$

15) $(5 - 8x)(-3) =$

16) $(-12)(3x + 3) =$

17) $(5 - 3x)6 =$

18) $4(2 + 6x) =$

19) $8(7x - 3) =$

20) $(-2x + 3)4 =$

21) $(7 - 5x)(-9) =$

22) $(-10)(x - 8) =$

23) $(11 - 4x)3 =$

24) $(-6)(10x - 4) =$

25) $(3 - 9x)(-7) =$

26) $(-9)(x + 9) =$

27) $(-3 + 5x)(-7) =$

28) $(-5)(8 - 10x) =$

29) $12(4x - 8) =$

30) $(-10x + 13)(-3) =$

31) $(-8)(3x - 2) + 4(x + 5) =$

32) $(-8)(x + 4) - (6 + 5x) =$

WWW.MathNotion.Com

ACT Aspire Subject Test Mathematics Grade 7

Evaluating One Variable Expressions

✎ Evaluate each expression using the value given.

1) $8 - x, x = 5$

2) $x - 9, x = 5$

3) $5x + 4, x = 3$

4) $x - 13, x = -4$

5) $12 - x, x = 4$

6) $x + 2, x = 6$

7) $4x + 8, x = 3$

8) $x + (-7), x = -8$

9) $4x + 5, x = 2$

10) $3x + 9, x = -2$

11) $15 + 3x - 7, x = 2$

12) $17 - 3x, x = 3$

13) $8x - 9, x = 4$

14) $5x + 4, x = -3$

15) $10x + 5, x = 3$

16) $14 - 4x, x = -6$

17) $3(5x + 3), x = 9$

18) $4(-3x - 6), x = 3$

19) $7x - 2x + 12, x = 4$

20) $(5x + 6) \div 2, x = 8$

21) $(x + 18) \div 10, x = 12$

22) $5x - 12 + 3x, x = -3$

23) $(6 - 4x)(-3), x = -4$

24) $9x^2 + 3x - 6, x = 2$

25) $x^2 - 10x, x = -5$

26) $3x(7 - 2x), x = 2$

27) $12x + 6 - 2x^2, x = -4$

28) $(-3)(4x - 8 + 3x), x = 3$

29) $(-6) + \frac{x}{4} + 3x, x = 16$

30) $(-6) + \frac{x}{5}, x = 35$

31) $\left(-\frac{45}{x}\right) - 7 + 2x, x = 9$

32) $\left(-\frac{21}{x}\right) - 12 + 4x, x = 7$

WWW.MathNotion.Com

ACT Aspire Subject Test Mathematics Grade 7

Evaluating Two Variables Expressions

✏️ Evaluate each expression using the values given.

1) $2x - 4y$,
   $x = 4, y = 1$

2) $3x + 5y$,
   $x = -2, y = 2$

3) $-7a + 4b$,
   $a = 2, b = 4$

4) $3x + 5 - y$,
   $x = 5, y = 6$

5) $3z + 12 - 2k$,
   $z = 5, k = 6$

6) $6(-x - 3y)$,
   $x = 5, y = -2$

7) $5a + 3b$,
   $a = 3, b = 4$

8) $7x \div 3y$,
   $x = 3, y = 7$

9) $2x + 15 + 5y$,
   $x = -3, y = 1$

10) $5a - (18 - b)$,
    $a = 2, b = 8$

11) $2z + 20 + 5k$,
    $z = -6, k = 5$

12) $xy + 10 + 4x$,
    $x = 3, y = 5$

13) $2x + 4y - 8 + 5$,
    $x = 5, y = 2$

14) $\left(-\frac{24}{x}\right) + 3 + 2y$,
    $x = 4, y = 6$

15) $(-3)(-3a - 3b)$,
    $a = 4, b = 5$

16) $12 + 4x - 7 - y$,
    $x = 3, y = 5$

17) $11x + 5 - 8y + 6$,
    $x = 5, y = 2$

18) $10 + 2(-4x - 5y)$,
    $x = 5, y = 4$

19) $5x + 13 + 6y$,
    $x = 5, y = 6$

20) $10a - (7a + 3b) - 11$,
    $a = 3, b = 8$

WWW.MathNotion.Com

ACT Aspire Subject Test Mathematics Grade 7

Combining like Terms

✎ **Simplify each expression.**

1) $11x + 3x + 6 =$

2) $8(2x - 6) =$

3) $18x - 7x + 11 =$

4) $(-4)(6x - 7) =$

5) $22x - 10x - 5 =$

6) $32x - 13 + 8x =$

7) $15 - (8x - 11) =$

8) $-24x + 17 - 11x =$

9) $12x - 8 - 6x + 9 =$

10) $21x + 5 - 36 + 12x =$

11) $28x + 3x - 11 =$

12) $(-3x + 4)5 =$

13) $2 + 4x + 9x - 8 =$

14) $6(2x - 5x) - 4 =$

15) $4(5x + 11) + 3x =$

16) $x - 14 - 11x =$

17) $5(10 + 9x) - 8x =$

18) $42x + 17 - 23x =$

19) $(-7x) + 19 + 20x =$

20) $(-7x) - 33 + 29x =$

21) $4(5x + 3) - 19x =$

22) $5(6 - 2x) - 15x =$

23) $-24x + (11 - 18x) =$

24) $(-9) - (6)(7x + 3) =$

25) $(-1)(8x - 10) - 21x =$

26) $-36x + 14 + 27x - 5x =$

27) $3(-13x + 6) - 17x =$

28) $-5x - 42 + 32x =$

29) $37x - 19x + 15 - 9x =$

30) $3(5x + 7x) - 31 =$

31) $14 - 6x - 15 - 9x =$

32) $-2(-5x - 7x) + 27x =$

ACT Aspire Subject Test Mathematics Grade 7

Answers of Worksheets

Translate Phrases into an Algebraic Statement

1) $9x$
2) $y - 11$
3) $\frac{19}{x}$
4) $38 - y$
5) $y + 40$
6) 6^2
7) x^5
8) $6 + x$
9) $57 - y$
10) $\frac{9}{x}$
11) $\frac{x^2}{25}$
12) $x - 6 = 19$
13) $10a - b^2$
14) $41 - ab$

Simplifying Variable Expressions

1) $3x + 15$
2) $-28x + 20$
3) $5x + 5$
4) $-8x^2 - 4$
5) $13x^2 + 10$
6) $18x^2 + 7x$
7) $-9x^2 + 4x$
8) $4x^2 - 10x$
9) $-22x + 21$
10) $68x - 12$
11) $-18x - 71$
12) $-11x^2 + 5x$
13) $-6x + 12$
14) $x - 4$
15) $22x - 5$
16) $14x - 22$
17) $27x - 16$
18) $25x + 28$
19) $-23x - 20$
20) $-3x^2 - 14x$
21) $-20x^2 + 27x$
22) $-20x^2 - 15x$
23) $4x^2 + 25x - 19$
24) $6x^2 - 66x + 25$
25) $3x^2 + 19x - 5$
26) $-7x^2 - 20x$
27) $-12x^2 + 10x - 7$
28) $-6x^2 - 21x + 13$
29) $10x^2 + 7x + 17$
30) $25x^2 + 25x$
31) $-16x^2 - 23x + 29$
32) $-9x^2 + 3x + 30$

The Distributive Property

1) $8x + 4$
2) $14x + 8$
3) $12x - 12$
4) $-12x + 30$
5) $-3x - 18$
6) $6x + 8$
7) $15x - 40$
8) $7x + 5$
9) $18x - 9$
10) $-4x + 28$
11) $3x - 5$
12) $12x + 27$
13) $18x + 24$
14) $-10x + 6$
15) $24x - 15$
16) $-36x - 36$
17) $-18x + 30$
18) $24x + 8$
19) $56x - 24$
20) $8x + 12$
21) $45x - 63$
22) $-10x + 80$
23) $-12x + 33$
24) $-60x + 24$
25) $63x - 21$
26) $-9x - 81$
27) $-35x + 21$
28) $50x - 40$
29) $48x - 96$
30) $30x - 39$
31) $-20x + 36$
32) $-13x - 38$

Evaluating One Variables

1) 3
2) -4
3) 19
4) -17
5) 8
6) 8
7) 20
8) -15
9) 13
10) 3
11) 14
12) 8

WWW.MathNotion.Com

ACT Aspire Subject Test Mathematics Grade 7

13) 23	18) −60	23) −66	28) −39
14) −11	19) 32	24) 36	29) 46
15) 35	20) 23	25) 75	30) 1
16) 38	21) 3	26) 18	31) 6
17) 144	22) −36	27) −74	32) 13

Evaluating Two Variables

1) 4	6) 6	11) 33	16) 12
2) 4	7) 27	12) 37	17) 50
3) 2	8) 1	13) 15	18) −70
4) 14	9) 14	14) 9	19) 74
5) 15	10) 0	15) 81	20) −26

Combining like Terms

1) $14x + 6$	9) $6x + 1$	17) $37x + 50$	25) $-29x + 10$
2) $16x - 48$	10) $33x - 31$	18) $19x + 17$	26) $-14x + 14$
3) $11x + 11$	11) $31x - 11$	19) $13x + 19$	27) $-56x + 18$
4) $-24x + 28$	12) $-15x + 20$	20) $22x - 33$	28) $27x - 42$
5) $12x - 5$	13) $13x - 6$	21) $x + 12$	29) $9x + 15$
6) $40x - 13$	14) $-18x - 4$	22) $-25x + 30$	30) $36x - 31$
7) $-8x + 26$	15) $23x + 44$	23) $-42x + 11$	31) $-15x - 1$
8) $-35x + 17$	16) $-10x - 14$	24) $-42x - 27$	32) $51x$

ACT Aspire Subject Test Mathematics Grade 7

Chapter 7 :
Equations and Inequalities

Topics that you will practice in this chapter:

- ✓ One–Step Equations
- ✓ Multi–Step Equations
- ✓ Graphing Single–Variable Inequalities
- ✓ One–Step Inequalities
- ✓ Multi-Step Inequalities
- ✓ Systems of Equations
- ✓ Systems of Equations Word Problems

"Life is a math equation. In order to gain the most, you have to know how to convert negatives into positives." – Anonymous

ACT Aspire Subject Test Mathematics Grade 7

One–Step Equations

✒ **Find the answer for each equation.**

1) $3x = 90, x = $ ___

2) $5x = 35, x = $ ___

3) $6x = 24, x = $ ___

4) $24x = 144, x = $ ___

5) $x + 15 = 20, x = $ ___

6) $x - 7 = 4, x = $ ___

7) $x - 9 = 2, x = $ ___

8) $x + 15 = 23, x = $ ___

9) $x - 4 = 13, x = $ ___

10) $12 = 16 + x, x = $ ___

11) $x - 10 = 2, x = $ ___

12) $5 - x = -11, x = $ ___

13) $28 = -6 + x, x = $ ___

14) $x - 20 = -35, x = $ ___

15) $x + 14 = -4, x = $ ___

16) $14 = 28 - x, x = $ ___

17) $7 + x = -7, x = $ ___

18) $x - 16 = 4, x = $ ___

19) $30 = x - 15, x = $ ___

20) $x - 5 = -18, x = $ ___

21) $x - 10 = 24, x = $ ___

22) $x - 20 = -25, x = $ ___

23) $x - 17 = 30, x = $ ___

24) $-70 = x - 28, x = $ ___

25) $x - 9 = 13, x = $ ___

26) $36 = 4x, x = $ ___

27) $x - 35 = 25, x = $ ___

28) $x - 25 = 10, x = $ ___

29) $70 - x = 16, x = $ ___

30) $x - 10 = 14, x = $ ___

31) $17 - x = -13, x = $ ___

32) $x - 9 = -30, x = $ ___

WWW.MathNotion.Com

ACT Aspire Subject Test Mathematics Grade 7

Multi–Step Equations

✏️ Find the answer for each equation.

1) $3x + 3 = 9$

2) $-x + 5 = 12$

3) $4x - 8 = 8$

4) $-(3 - x) = 5$

5) $4x - 8 = 16$

6) $12x - 15 = 9$

7) $2x - 18 = 2$

8) $4x + 8 = 16$

9) $24x + 27 = 75$

10) $-14(3 + x) = 14$

11) $-3(2 + x) = 6$

12) $12 = -(x - 7)$

13) $3(3 - x) = 30$

14) $-15 = -(3x + 6)$

15) $40(3 + x) = 40$

16) $5(x - 10) = 25$

17) $-18 = x + 8x$

18) $3x + 25 = -2x - 10$

19) $7(6 + 3x) = -63$

20) $18 - 3x = -4 - 5x$

21) $4 - 6x = 36 + 2x$

22) $15 + 15x = -5 + 5x$

23) $42 = (-6x) - 7 + 7$

24) $21 = 3x - 21 + 4x$

25) $-18 = -6x - 9 + 3x$

26) $5x - 15 = -29 + 6x$

27) $7x - 18 = 4x + 3$

28) $-7 - 4x = 5(4 - x)$

29) $x - 5 = -5(-3 - x)$

30) $13x - 68 = 15x - 102$

31) $-5x - 3 = -3(9 + 3x)$

32) $-2x - 15 = 6x + 17$

ACT Aspire Subject Test Mathematics Grade 7

Graphing Single–Variable Inequalities

✎ Draw a graph for each inequality.

1) $x > -1$

2) $x \leq 2$

3) $x \geq 0$

4) $x < -3$

5) $x < \frac{1}{2}$

6) $x \leq -2$

7) $x \leq 3$

8) $x \geq -\frac{7}{2}$

ACT Aspire Subject Test Mathematics Grade 7

One–Step Inequalities

 Find the answer for each inequality and graph it.

1) $x + 4 \geq 4$

2) $x - 5 \leq 2$

3) $5x > 35$

4) $9 + x \leq 11$

5) $x - 5 < -9$

6) $9x \geq 72$

7) $9x \leq 27$

8) $x + 19 > 16$

WWW.MathNotion.Com

ACT Aspire Subject Test Mathematics Grade 7

Multi-Step Inequalities

✎ **Calculate each inequality.**

1) $x - 3 \leq 7$

2) $8 - x \leq 8$

3) $3x - 9 \leq 9$

4) $4x - 4 \geq 8$

5) $x - 7 \geq 1$

6) $5x - 15 \leq 5$

7) $6x - 8 \leq 4$

8) $-11 + 6x \leq 12$

9) $4(x - 4) \leq 16$

10) $3x - 10 \leq 11$

11) $5x - 25 < 25$

12) $9x - 5 < 22$

13) $20 - 7x \geq -15$

14) $33 + 6x < 45$

15) $8 + 8x \geq 96$

16) $7 + 3x < 13$

17) $4x - 3 < 9$

18) $5(2 - 2x) \geq -30$

19) $-(7 + 6x) < 29$

20) $12 - 8x \geq -20$

21) $-4(x - 6) > 24$

22) $\dfrac{3x + 9}{6} \leq 10$

23) $\dfrac{4x - 10}{3} \leq 2$

24) $\dfrac{2x - 8}{3} > 2$

25) $8 + \dfrac{x}{6} < 9$

26) $\dfrac{9x}{7} - 4 < 5$

27) $\dfrac{15x + 45}{15} > 1$

28) $16 + \dfrac{x}{4} < 6$

WWW.MathNotion.Com

ACT Aspire Subject Test Mathematics Grade 7

Systems of Equations

✏️ **Calculate each system of equations.**

1) $-x + y = 2$ $x =$ ___
   $-4x + 2y = 6$ $y =$ ___

2) $-15x + 3y = -9$ $x =$ ___
   $9x - 16y = 48$ $y =$ ___

3) $y = -7$ $x =$ ___
   $6x + 5y = 7$ $y =$ ___

4) $3y = -9x + 15$ $x =$ ___
   $5x - 4y = -3$ $y =$ ___

5) $10x - 9y = -13$ $x =$ ___
   $-5x + 3y = 11$ $y =$ ___

6) $-12x - 16y = 20$ $x =$ ___
   $6x - 12y = 30$ $y =$ ___

7) $5x - 14y = -23$ $x =$ ___
   $-18x + 21y = 24$ $y =$ ___

8) $15x - 21y = -6$ $x =$ ___
   $2x - 3y = -2$ $y =$ ___

9) $-x + 3y = 3$ $x =$ ___
   $-14x + 16y = -10$ $y =$ ___

10) $x + 5y = 50$ $x =$ ___
    $3x + 10y = 80$ $y =$ ___

11) $6x - 7y = -8$ $x =$ ___
    $-x - 4y = -9$ $y =$ ___

12) $2x + 4y = -10$ $x =$ ___
    $2x - 8y = 14$ $y =$ ___

13) $4x + 3y = 12$ $x =$ ___
    $5x - 3y = 15$ $y =$ ___

14) $3x - 2y = 3$ $x =$ ___
    $7x - 8y = 22$ $y =$ ___

15) $3x + 2y = 5$ $x =$ ___
    $-10x - 4y = -14$ $y =$ ___

16) $10x + 7y = 1$ $x =$ ___
    $-5x - 7y = 24$ $y =$ ___

WWW.MathNotion.Com

ACT Aspire Subject Test Mathematics Grade 7

Systems of Equations Word Problems

✎ Find the answer for each word problem.

1) Tickets to a movie cost $4 for adults and $3 for students. A group of friends purchased 8 tickets for $31.00. How many adults ticket did they buy? ____

2) At a store, Eva bought two shirts and five hats for $77.00. Nicole bought three same shirts and four same hats for $84.00. What is the price of each shirt? ____

3) A farmhouse shelters 18 animals, some are pigs, and some are ducks. Altogether there are 66 legs. How many pigs are there? ____

4) A class of 214 students went on a field trip. They took 36 vehicles, some cars and some buses. If each car holds 5 students and each bus hold 22 students, how many buses did they take? ____

5) A theater is selling tickets for a performance. Mr. Smith purchased 5 senior tickets and 3 child tickets for $105 for his friends and family. Mr. Jackson purchased 3 senior tickets and 5 child tickets for $79. What is the price of a senior ticket? $____

6) The difference of two numbers is 10. Their sum is 20. What is the bigger number? $____

7) The sum of the digits of a certain two-digit number is 7. Reversing its digits increase the number by 9. What is the number? ____

8) The difference of two numbers is 11. Their sum is 25. What are the numbers? _____

9) The length of a rectangle is 5 meters greater than 2 times the width. The perimeter of rectangle is 28 meters. What is the length of the rectangle? _____

10) Jim has 25 nickels and dimes totaling $1.80. How many nickels does he have? ____

WWW.MathNotion.Com

ACT Aspire Subject Test Mathematics Grade 7

Answers of Worksheets

One–Step Equations

1) 30	9) 17	17) −14	25) 22
2) 7	10) −4	18) 20	26) 9
3) 4	11) 12	19) 45	27) 60
4) 6	12) 16	20) −13	28) 35
5) 5	13) 34	21) 34	29) 54
6) 11	14) −15	22) −5	30) 24
7) 11	15) −18	23) 47	31) 30
8) 8	16) 14	24) −42	32) −21

Multi–Step Equations

1) 2	9) 2	17) −2	25) 3
2) −7	10) −4	18) −7	26) 14
3) 4	11) −4	19) −5	27) 7
4) 8	12) −5	20) −11	28) 27
5) 6	13) −7	21) −4	29) −5
6) 2	14) 3	22) −2	30) 17
7) 10	15) −2	23) −7	31) −6
8) 2	16) 15	24) 6	32) −4

Graphing Single–Variable Inequalities

1)

2)

3)

4)

ACT Aspire Subject Test Mathematics Grade 7

5)
6)
7)
8)

One–Step Inequalities

1)
2)
3)
4)
5)
6)
7)
8)

Multi-Step Inequalities

1) $x \leq 10$
2) $x \geq 0$
3) $x \leq 6$
4) $x \geq 3$
5) $x \geq 8$
6) $x \leq 4$
7) $x \leq 2$
8) $x \leq \frac{23}{6}$
9) $x \leq 8$
10) $x \leq 7$
11) $x < 10$
12) $x < 3$
13) $x \leq 5$
14) $x < 2$
15) $x \geq 11$
16) $x < 2$
17) $x < 3$
18) $x \leq 4$
19) $x > -6$
20) $x \leq 4$
21) $x < 0$
22) $x \leq 17$
23) $x \leq 4$
24) $x > 7$

ACT Aspire Subject Test Mathematics Grade 7

25) $x < 6$ 26) $x < 7$ 27) $x > -2$ 28) $x < -40$

Systems of Equations

1) $x = -1, y = 1$
2) $x = 0, y = -3$
3) $x = 7$
4) $x = 1, y = 2$
5) $x = -4, y = -3$
6) $x = 1, y = -2$
7) $x = 1, y = 2$
8) $x = 8, y = 6$
9) $x = 3, y = 2$
10) $x = -20, y = 14$
11) $x = 1, y = 2$
12) $x = -1, y = -2$
13) $x = 3, y = 0$
14) $x = -2, y = -\frac{9}{2}$
15) $x = 1, y = 1$
16) $x = 5, y = -7$

Systems of Equations Word Problems

1) 7
2) $16
3) 15
4) 2
5) $18
6) 15
7) 34
8) 18, 7
9) 11 meters
10) 14

WWW.MathNotion.Com

Chapter 8:
Linear Functions

Topics that you will practice in this chapter:

- ✓ Finding Slope
- ✓ Graphing Lines Using Line Equation
- ✓ Writing Linear Equations
- ✓ Graphing Linear Inequalities
- ✓ Finding Midpoint
- ✓ Finding Distance of Two Points

"Nature is written in mathematical language." – Galileo Galilei

ACT Aspire Subject Test Mathematics Grade 7

Finding Slope

👉 Find the slope of each line.

1) $y = x + 8$

2) $y = -3x + 5$

3) $y = 2x + 12$

4) $y = -4x + 19$

5) $y = 11 + 6x$

6) $y = 7 - 5x$

7) $y = 8x + 19$

8) $y = -9x + 20$

9) $y = -7x + 4$

10) $y = 3x - 8$

11) $y = \frac{1}{3}x + 8$

12) $y = -\frac{4}{5}x + 9$

13) $-3x + 6y = 30$

14) $4x + 4y = 16$

15) $3y - x = 10$

16) $8y - x = 5$

👉 Find the slope of the line through each pair of points.

17) $(2, 3), (7, 10)$

18) $(-3, 5), (2, 15)$

19) $(5, -3), (1, 9)$

20) $(-5, -5), (10, 25)$

21) $(22, 3), (7, 18)$

22) $(-16, 8), (-7, 26)$

23) $(25, 11), (29, 19)$

24) $(26, -19), (14, 17)$

25) $(22, -13), (20, -11)$

26) $(19, 7), (15, -3)$

27) $(5, 7), (11, 19)$

28) $(52, -62), (40, 70)$

WWW.MathNotion.Com

ACT Aspire Subject Test Mathematics Grade 7

Graphing Lines Using Line Equation

✏️ **Sketch the graph of each line.**

1) $y = x - 2$

2) $y = -3x + 2$

3) $x + y = 0$

  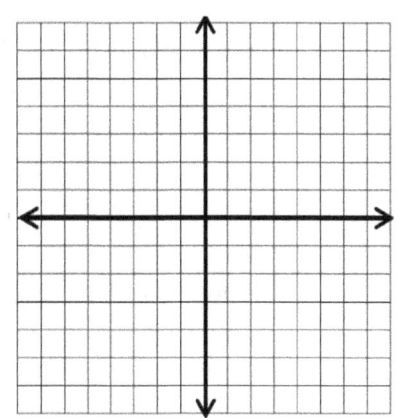

4) $x + y = -3$

5) $2x + 3y = -4$

6) $y - 3x + 6 = 0$

  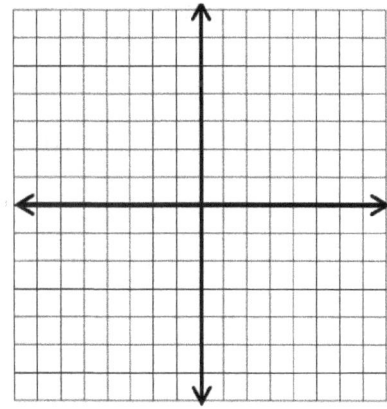

ACT Aspire Subject Test Mathematics Grade 7

Writing Linear Equations

✍ **Write the equation of the line through the given points.**

1) Through: $(2, -5), (3, 9)$

2) Through: $(-6, 3), (3, 12)$

3) Through: $(10, 7), (5, 27)$

4) Through: $(15, 11), (3, -1)$

5) Through: $(24, 17), (12, -7)$

6) Through: $(8, 29), (4, -7)$

7) Through: $(20, -16), (12, 0)$

8) Through: $(-3, 10), (2, -5)$

9) Through: $(-6, 17), (4, -3)$

10) Through: $(-8, 22), (5, -4)$

11) Through: $(9, 27), (3, -3)$

12) Through: $(11, 32), (9, 4)$

13) Through: $(-3, 13), (-4, 0)$

14) Through: $(-5, 5), (5, 15)$

15) Through: $(18, -32), (11, 3)$

16) Through: $(-4, 25), (4, -15)$

✍ **Find the answer for each problem.**

17) What is the equation of a line with slope 6 and intercept 12?
    _____

18) What is the equation of a line with slope -11 and intercept -4?
    _____

19) What is the equation of a line with slope -3 and passes through point $(5, 2)$? _____

20) What is the equation of a line with slope -5 and passes through point $(-2, -1)$? _____

21) The slope of a line is -10 and it passes through point $(-3, 0)$. What is the equation of the line? _____

22) The slope of a line is 8 and it passes through point $(0, 7)$. What is the equation of the line? _____

Graphing Linear Inequalities

✎ **Sketch the graph of each linear inequality.**

1) $y > 4x - 5$

2) $y < 2x + 4$

3) $y \leq -5x - 2$

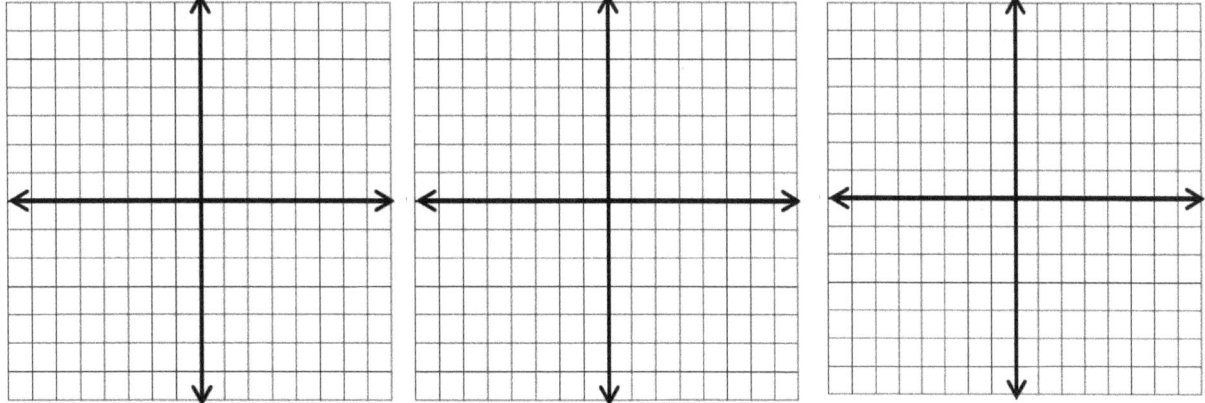

4) $4y \geq 12 + 4x$

5) $-12y < 3x - 24$

6) $5y \geq -15x + 10$

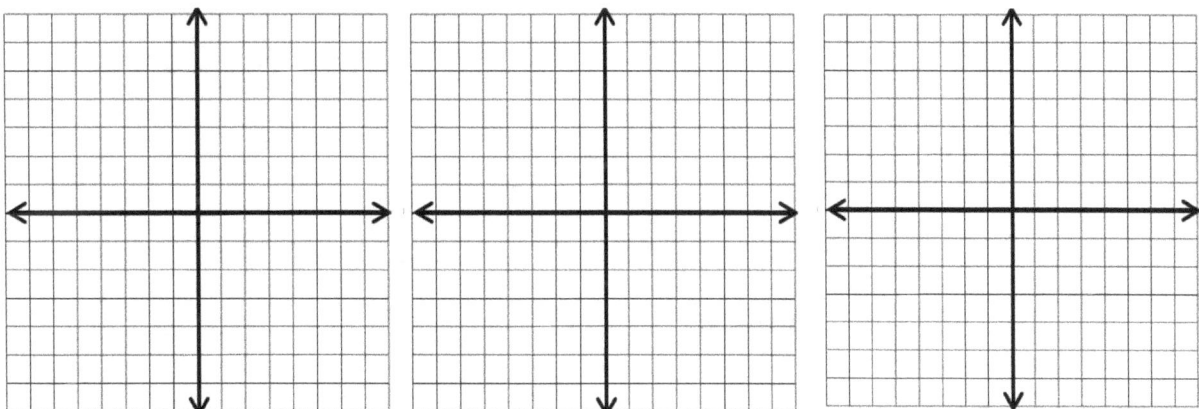

ACT Aspire Subject Test Mathematics Grade 7

Finding Midpoint

✎ **Find the midpoint of the line segment with the given endpoints.**

1) $(-4, -3), (2, 3)$

2) $(9, 0), (-1, 8)$

3) $(9, -6), (3, 14)$

4) $(-10, -6), (0, 8)$

5) $(2, -5), (14, -15)$

6) $(-10, -3), (4, -13)$

7) $(8, 7), (-8, 13)$

8) $(-3, 6), (-9, 2)$

9) $(-4, 5), (16, -9)$

10) $(7, 14), (9, -2)$

11) $(-8, 6), (6, 6)$

12) $(10, 5), (-2, -3)$

13) $(-5, 12), (-3, 3)$

14) $(12, 7), (8, -2)$

15) $(10, 2), (-6, 14)$

16) $(-1, -2), (-7, 10)$

17) $(7, -7), (13, -13)$

18) $(-3, -8), (11, -4)$

19) $(5, -11), (-8, 9)$

20) $(14, -4), (16, 14)$

21) $(0, -5), (8, -1)$

22) $(3, 0), (-21, 18)$

23) $(17, -3), (-7, -5)$

24) $(26, -12), (6, 24)$

✎ **Find the answer for each problem.**

25) One endpoint of a line segment is $(-3, 7)$ and the midpoint of the line segment is $(-6, 9)$. What is the other endpoint? _____

26) One endpoint of a line segment is $(-3, 7)$ and the midpoint of the line segment is $(1, 5)$. What is the other endpoint? _____

27) One endpoint of a line segment is $(-10, -16)$ and the midpoint of the line segment is $(2, 9)$. What is the other endpoint? _____

ACT Aspire Subject Test Mathematics Grade 7

Finding Distance of Two Points

✎ Find the distance between each pair of points.

1) $(6, 3), (-3, -9)$

2) $(5, 2), (-10, -6)$

3) $(8, 5), (8, 3)$

4) $(-8, -2), (2, 22)$

5) $(6, -7), (-3, -7)$

6) $(12, 0), (-9, -20)$

7) $(3, 20), (3, -5)$

8) $(10, 17), (5, 5)$

9) $(7, -2), (-4, -2)$

10) $(13, 4), (5, -2)$

11) $(11, 13), (5, 5)$

12) $(1, 4), (-23, -3)$

13) $(9, 8), (5, -4)$

14) $(-11, -4), (5, 8)$

15) $(-2, -6), (-2, -12)$

16) $(-1, -4), (23, 3)$

17) $(19, 3), (7, -6)$

18) $(-5, -2), (3, 4)$

19) $(2, 6), (2, -12)$

20) $(-4, -2), (8, -2)$

✎ Find the answer for each problem.

21) Triangle ABC is a right triangle on the coordinate system and its vertices are $(-2, 5), (-2, 1)$, and $(1, 1)$. What is the area of triangle ABC? _____

22) Three vertices of a triangle on a coordinate system are $(3, -6)$, $(-5, -12)$, and $(3, -18)$. What is the perimeter of the triangle? _____

23) Four vertices of a rectangle on a coordinate system are $(-2, 2), (-2, 6)$, $(4, 2)$, and $(4, 6)$. What is its perimeter? _____

ACT Aspire Subject Test Mathematics Grade 7

Answers of Worksheets

Finding Slope

1) 1
2) -3
3) 2
4) -4
5) 6
6) -5
7) 8
8) -9
9) -7
10) 3
11) $\frac{1}{3}$
12) $-\frac{4}{5}$
13) $\frac{1}{2}$
14) -1
15) $\frac{1}{3}$
16) $\frac{1}{8}$
17) $\frac{7}{5}$
18) 2
19) -3
20) 2
21) -1
22) 2
23) 2
24) -3
25) -1
26) $\frac{5}{2}$
27) 2
28) -11

Graphing Lines Using Line Equation

1) $y = x - 2$

2) $y = -3x + 2$

3) $x + y = 0$

4) $x + y = -3$

5) $2x + 3y = -4$

6) $y - 3x + 6 = 0$

ACT Aspire Subject Test Mathematics Grade 7

Writing Linear Equations

1) $y = 14x - 33$
2) $y = x + 9$
3) $y = -4x + 47$
4) $y = x - 4$
5) $y = 2x - 31$
6) $y = 9x - 43$
7) $y = -2x + 24$
8) $y = -3x + 1$
9) $y = -2x + 5$
10) $y = -2x + 6$
11) $y = 5x - 18$
12) $y = 14x - 122$
13) $y = 13x + 52$
14) $y = x + 10$
15) $y = -5x + 58$
16) $y = -5x + 5$
17) $y = 6x + 12$
18) $y = -11x - 4$
19) $y = -3x + 17$
20) $y = -5x - 11$
21) $y = -10x - 30$
22) $y = 8x + 7$

Graphing Linear Inequalities

1) $y > 4x - 5$

2) $y < 2x + 4$

3) $y \leq -5x - 2$

4) $4y \geq 12 + 4x$

5) $-12y < 3x - 24$

6) $5y \geq -15x + 10$

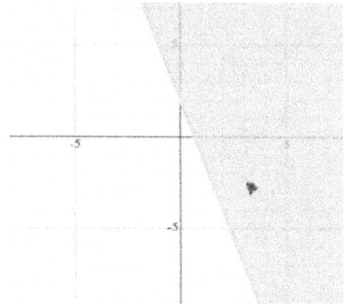

Finding Midpoint

1) $(-1, 0)$
2) $(4, 4)$
3) $(6, 4)$
4) $(-5, 1)$
5) $(8, -10)$
6) $(-3, -8)$
7) $(0, 10)$
8) $(-6, 4)$
9) $(6, -2)$
10) $(8, 6)$
11) $(-1, 6)$
12) $(4, 1)$
13) $(-4, 7.5)$
14) $(10, 2.5)$
15) $(2, 8)$
16) $(-4, 4)$
17) $(10, -10)$
18) $(4, -6)$

ACT Aspire Subject Test Mathematics Grade 7

19) (−1.5, −1)
20) (15, 5)
21) (4, −3)
22) (−9, 9)
23) (5, −4)
24) (16, 6)
25) (−9, 11)
26) (5, 3)
27) (14, 34)

Finding Distance of Two Points

1) 15
2) 17
3) 2
4) 26
5) 9
6) 29
7) 25
8) 13
9) 11
10) 10
11) 10
12) 25
13) $4\sqrt{10}$
14) 20
15) 6
16) 25
17) 15
18) 10
19) 18
20) 12
21) 6 *square units*
22) 32 *units*
23) 20 *units*

ACT Aspire Subject Test Mathematics Grade 7

Chapter 9 :
Transformations

Topics that you will learn in this chapter:

- ✓ Translations
- ✓ Reflections
- ✓ Rotations
- ✓ Dilations
- ✓ Coordinate of Vertices.

"Nothing gets transformed in your life until your mind is transformed."

Ifeanyi Enoch Onuoha

ACT Aspire Subject Test Mathematics Grade 7

Translations

✎ **Graph the image of the figure using the transformation given.**

1) Translation: 5 units right and 2 units down

2) Translation: 1 units left and 3 units down

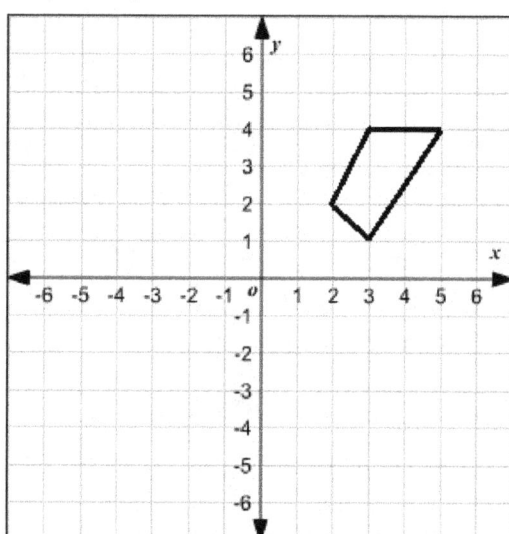

✎ **Write a rule to describe each transformation.**

3)

4)

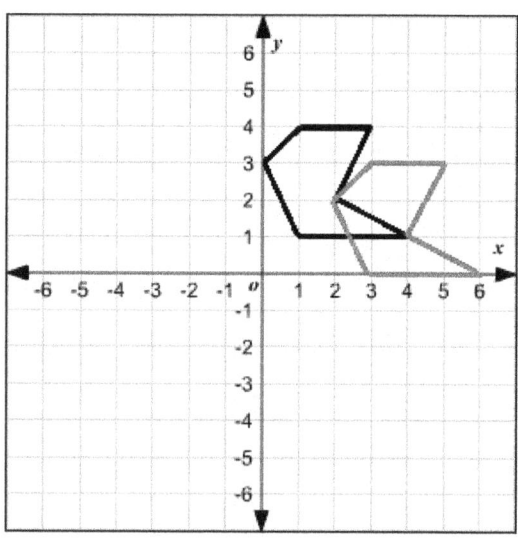

WWW.MathNotion.Com

ACT Aspire Subject Test Mathematics Grade 7

Reflections

✎ **Graph the image of the figure using the transformation given.**

1) Reflection across $x = 3$

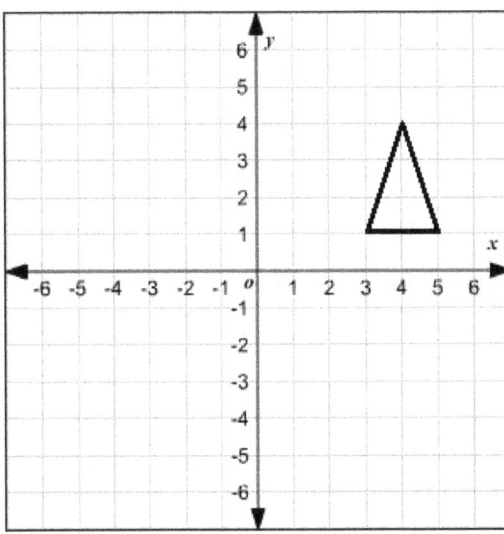

2) Reflection across $y = x$

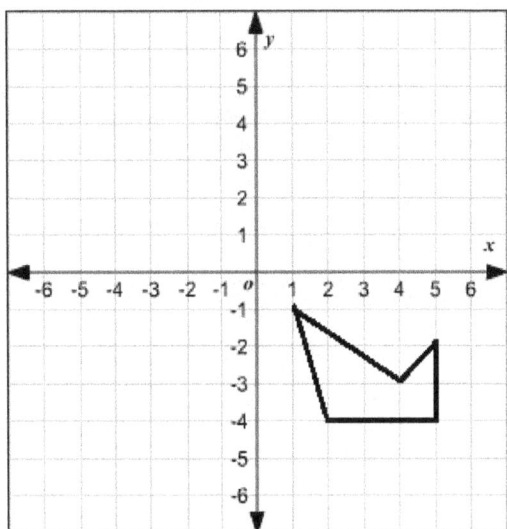

3) Reflection across $y = -2$

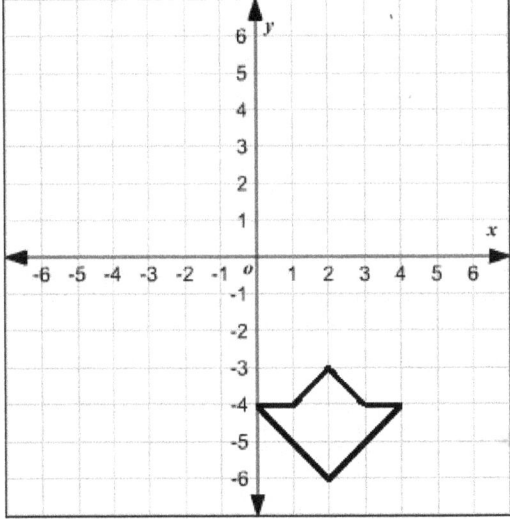

4) Reflection across y axis

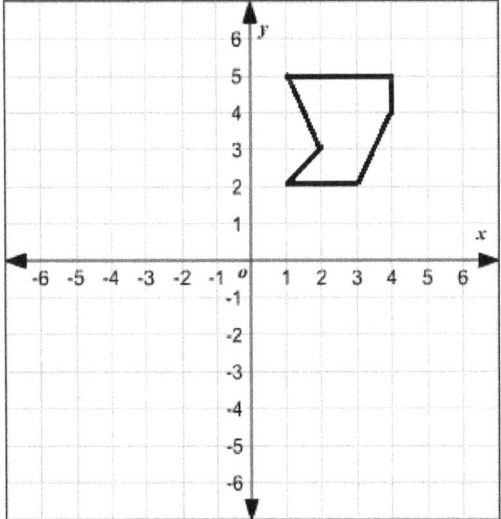

WWW.MathNotion.Com

ACT Aspire Subject Test Mathematics Grade 7

✎ **Write a rule to describe each transformation.**

5)

6)

7)

8)

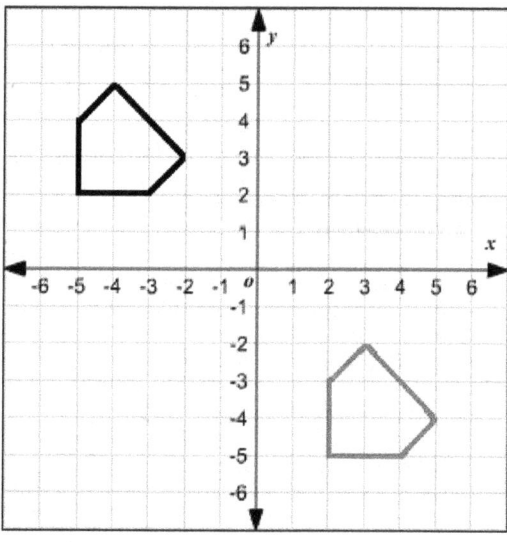

WWW.MathNotion.Com

ACT Aspire Subject Test Mathematics Grade 7

Rotations

✎ **Graph the image of the figure using the transformation given.**

1) Rotation 90° counterclockwise about the origin

2) Rotation 270° counterclockwise about the origin

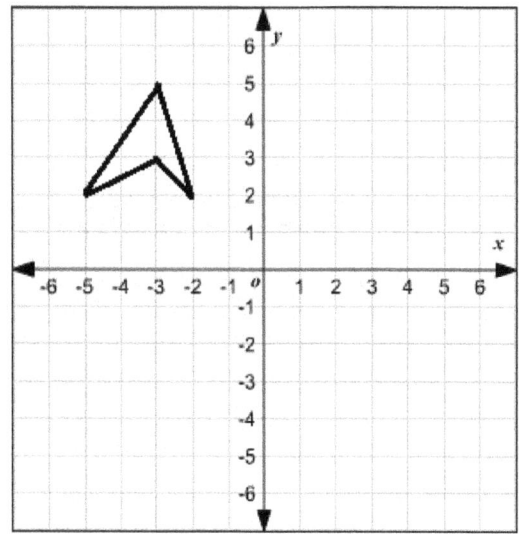

3) Rotation 270° clockwise about the origin

4) Rotation 180° counterclockwise about the origin

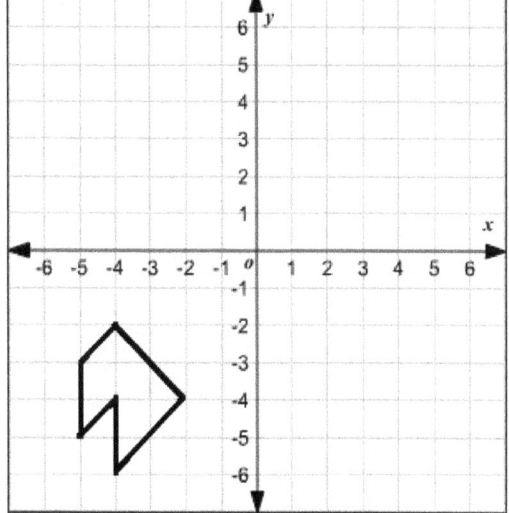

WWW.MathNotion.Com

ACT Aspire Subject Test Mathematics Grade 7

✏️ **Write a rule to describe each transformation.**

5)

6)

7)

8)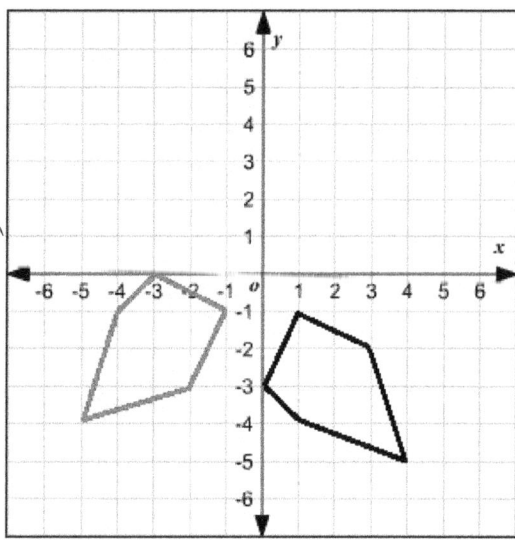

ACT Aspire Subject Test Mathematics Grade 7

Dilations

✏️ **Determine whether the dilation from figure M to figure N is a reduction or an enlargement. Then find the scale factor and the missing length.**

1)

2)
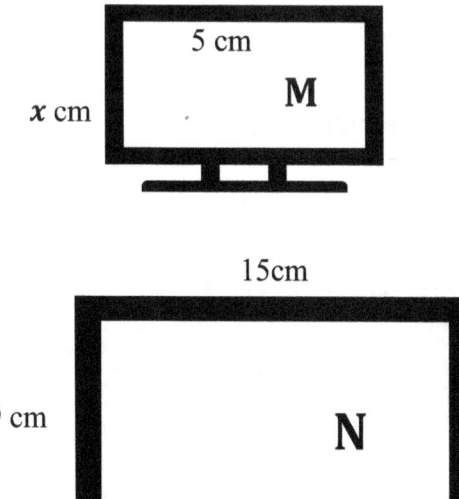

✏️ **Draw a dilation of the figure using the given scale factor.**

3) $k = \frac{1}{2}$

4) $k = 3$

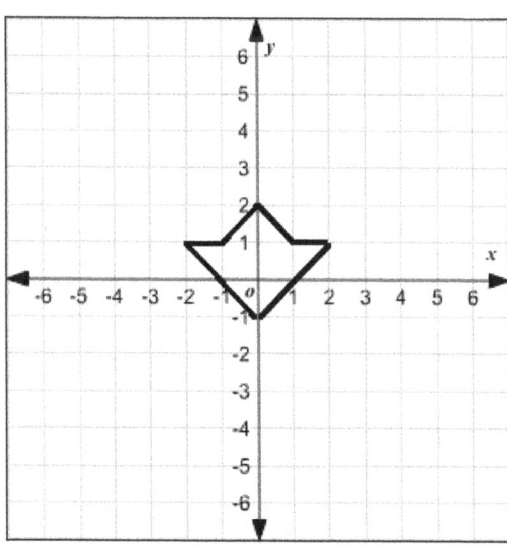

ACT Aspire Subject Test Mathematics Grade 7

Coordinates of Vertices

✎ **Calculate the new coordinates after the given transformations.**

1) Translate: 1-unit right and 2 units down.

   $A(-1, 0), B(1, -3), C(2, 1)$

2) Rotation: 270° clockwise about the origin.

   $D(-1, 1), E(-5, 6), F(-7, 3), G(-3, 10)$

3) Rotation: 180° counterclockwise about the origin.

   $P(3, 0), Q(5, 2), R(-1, 4), S(2, -7)$

4) Rotation: 90° clockwise about the origin.

   $J(-2, 5), K(-5, 0), L(3, -6)$

5) Reflection: over the y axis.

   $C(-2, -6), D(4, -3), W(1, -7), Y(5, 1)$

6) Reflection: across the line $y = -x$.

   $A(3, -2), B(8, -4), C(6, -6), D(1, -5)$

7) Reflection: across the line $y = -2$.

   $K(-1, 1), L(-4, 2), M(4, -1), N(2, 3)$

8) Dilate: Reduction by scale factor $\frac{1}{3}$.

   $A(6, 3), B(-12, 0), C(-9, 6)$

9) Dilate: Enlargement by scale factor 2.

   $F(-1, 4), G(-3, 0), H(3, 2)$

ACT Aspire Subject Test Mathematics Grade 7

Answers of Worksheets

Translations

1)

2)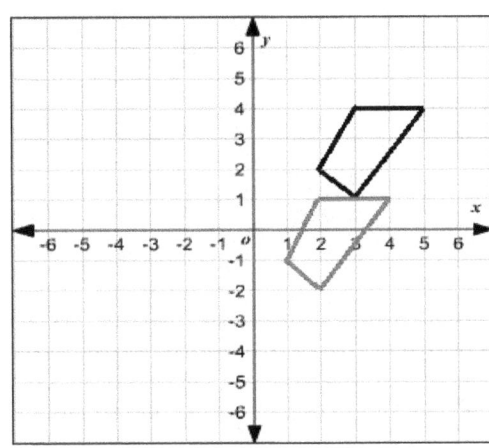

3) Translation: : 1 units left and 4 units down

4) Translation: 2 units right and 1 unit down

Reflections

1)

2)

3)

4)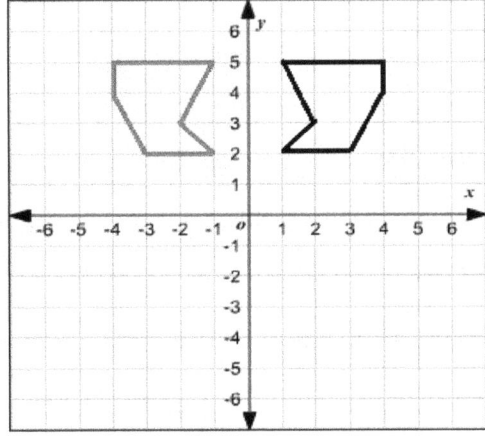

ACT Aspire Subject Test Mathematics Grade 7

5) Reflection across the x = 0 (y axis)

6) Reflection across the y = 1

7) Reflection against the origin

8) Reflection across the y = x

Rotations

1)

2)

3)

4)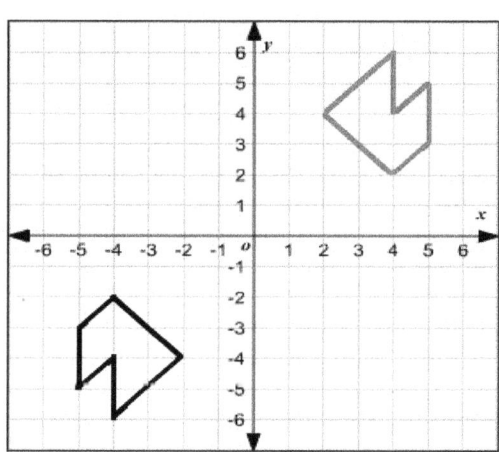

5) Rotation 90° clockwise about the origin

6) Rotation 180° about the origin

7) Rotation 270° counter clockwise about the origin

8) Rotation 90° clockwise about the origin

Dilations

1) Reduction, $k = \frac{3}{2}$, $x = 4\ cm$

2) Enlargement, $k = \frac{1}{3}$, $x = 3\ cm$

ACT Aspire Subject Test Mathematics Grade 7

3)

4)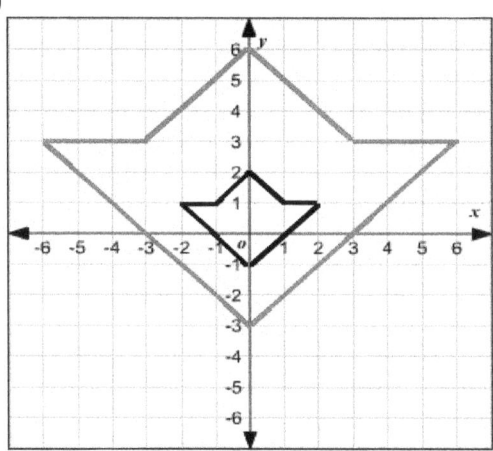

Coordinate of Vertices

1) $A'(0, -2), B'(2, -5), C'(3, -1)$
2) $D'(-1, -1), E'(-6, -5), F'(-3, -7), G'(-10, -3)$
3) $P'(-3, 0), Q'(-5, -2), R'(1, -4), S'(-2, 7)$
4) $J'(5, 2), K'(0, 5), L'(-6, -3)$
5) $C'(2, -6), D'(-4, -3), W'(-1, -7), Y'(-5, 1)$
6) $A'(2, -3), B'(4, -8), C'(6, -6), D'(5, -1)$
7) $K'(-1, -5), L'(-4, -6), M'(4, -3), N'(2, -7)$
8) $A'(2, 1), B'(-4, 0), C'(-3, 2)$
9) $F'(-2, 8), G'(-6, 0), H'(6, 4)$

ACT Aspire Subject Test Mathematics Grade 7

Chapter 10 :

Geometry and Solid Figures

Topics that you will practice in this chapter:

- ✓ Angles
- ✓ Pythagorean Relationship
- ✓ Triangles
- ✓ Polygons
- ✓ Trapezoids
- ✓ Circles
- ✓ Cubes
- ✓ Rectangular Prism
- ✓ Cylinder
- ✓ Pyramids and Cone

Mathematics is, as it were, a sensuous logic, and relates to philosophy as do the arts, music, and plastic art to poetry. — K. Shegel

Angles

What is the value of x in the following figures?

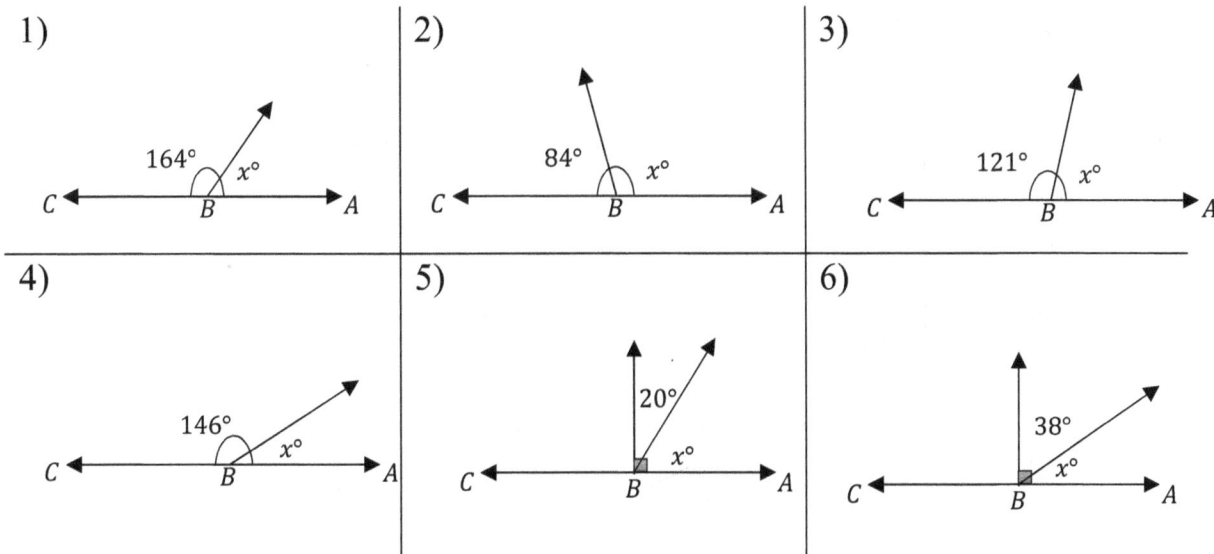

Calculate.

7) Two supplement angles have equal measures. What is the measure of each angle? _____

8) The measure of an angle is seven fifth the measure of its supplement. What is the measure of the angle? _____

9) Two angles are complementary and the measure of one angle is 24 less than the other. What is the measure of the smaller angle? _____

10) Two angles are complementary. The measure of one angle is one fifth the measure of the other. What is the measure of the bigger angle? _____

11) Two supplementary angles are given. The measure of one angle is 40° less than the measure of the other. What does the smaller angle measure? _____

Pythagorean Relationship

✎ **Do the following lengths form a right triangle?**

1)

2)

3)

4)

5)

6)

7)

8)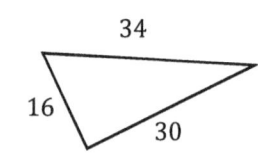

✎ **Find the missing side?**

9)

10)

11)

12)

13)

14)

15)

16)

ACT Aspire Subject Test Mathematics Grade 7

Triangles

✏️ **Find the measure of the unknown angle in each triangle.**

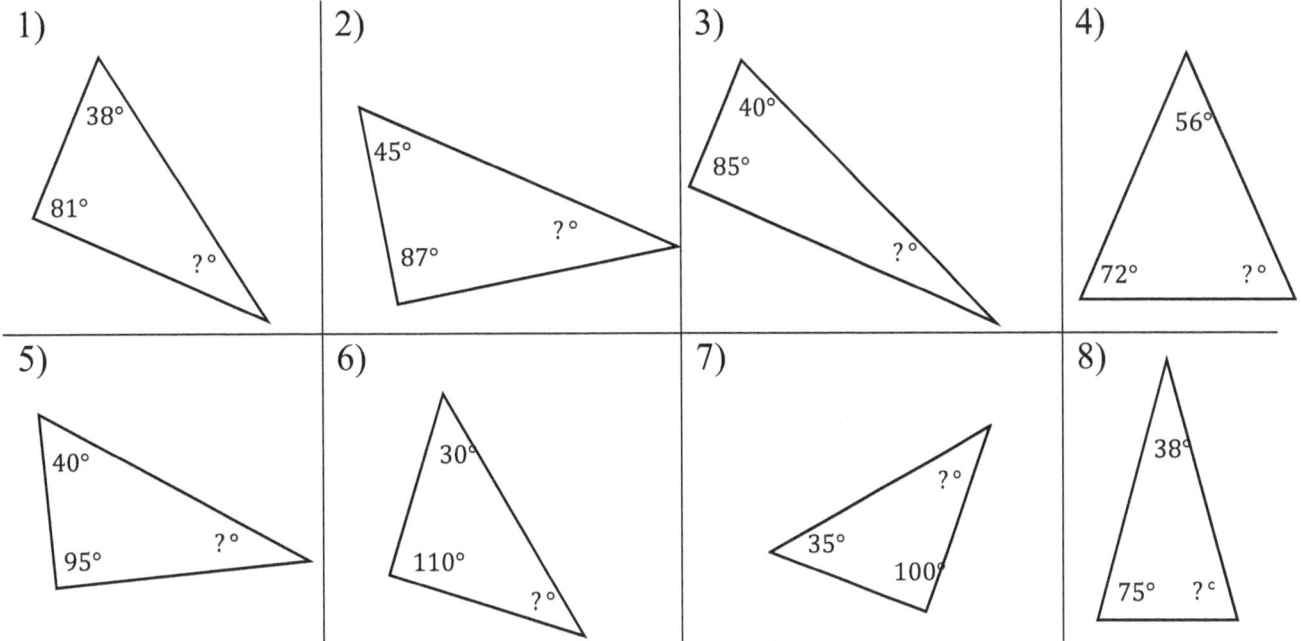

✏️ **Find area of each triangle.**

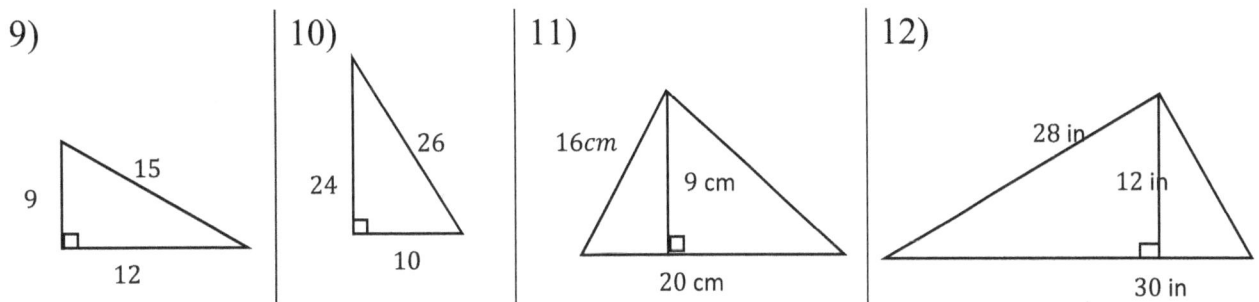

WWW.MathNotion.Com

ACT Aspire Subject Test Mathematics Grade 7

Polygons

✏️ **Find the perimeter of each shape.**

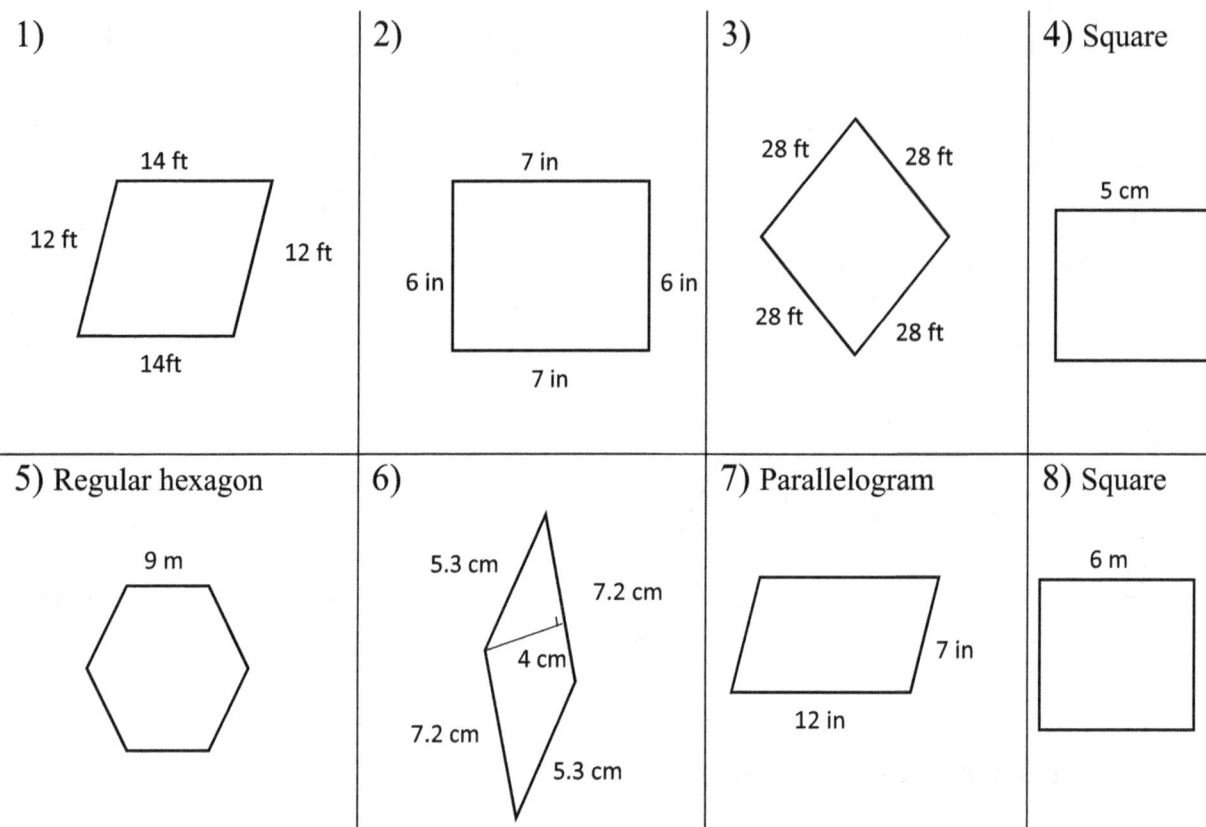

✏️ **Find the area of each shape.**

9) Parallelogram   10) Rectangle   11) Rectangle   12) Square

WWW.MathNotion.Com

ACT Aspire Subject Test Mathematics Grade 7

Trapezoids

✏️ **Find the area of each trapezoid.**

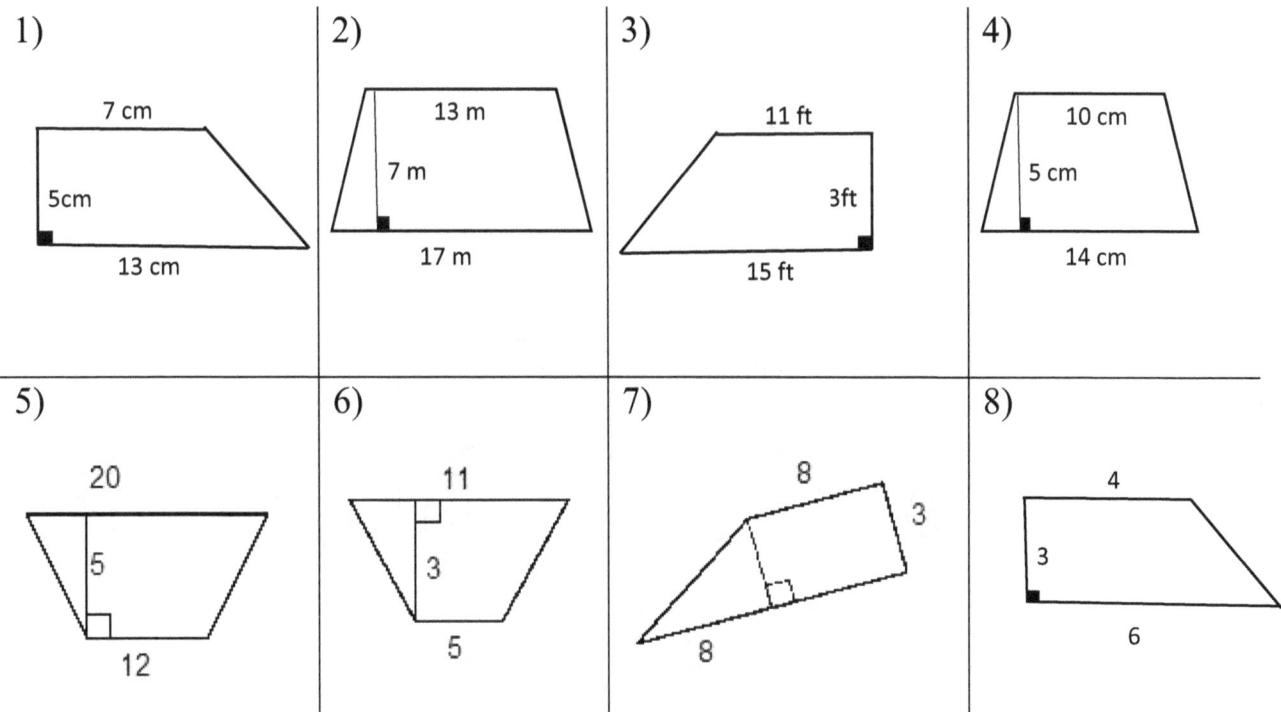

✏️ **Calculate.**

1) A trapezoid has an area of 45 cm² and its height is 5 cm and one base is 5 cm. What is the other base length? _____

2) If a trapezoid has an area of 99 ft² and the lengths of the bases are 8 ft and 10 ft, find the height? _____

3) If a trapezoid has an area of 126 m² and its height is 14 m and one base is 6 m, find the other base length? _____

4) The area of a trapezoid is 440 ft² and its height is 22 ft. If one base of the trapezoid is 15 ft, what is the other base length? _____

ACT Aspire Subject Test Mathematics Grade 7

Circles

✏️ **Find the area of each circle.** ($\pi = 3.14$)

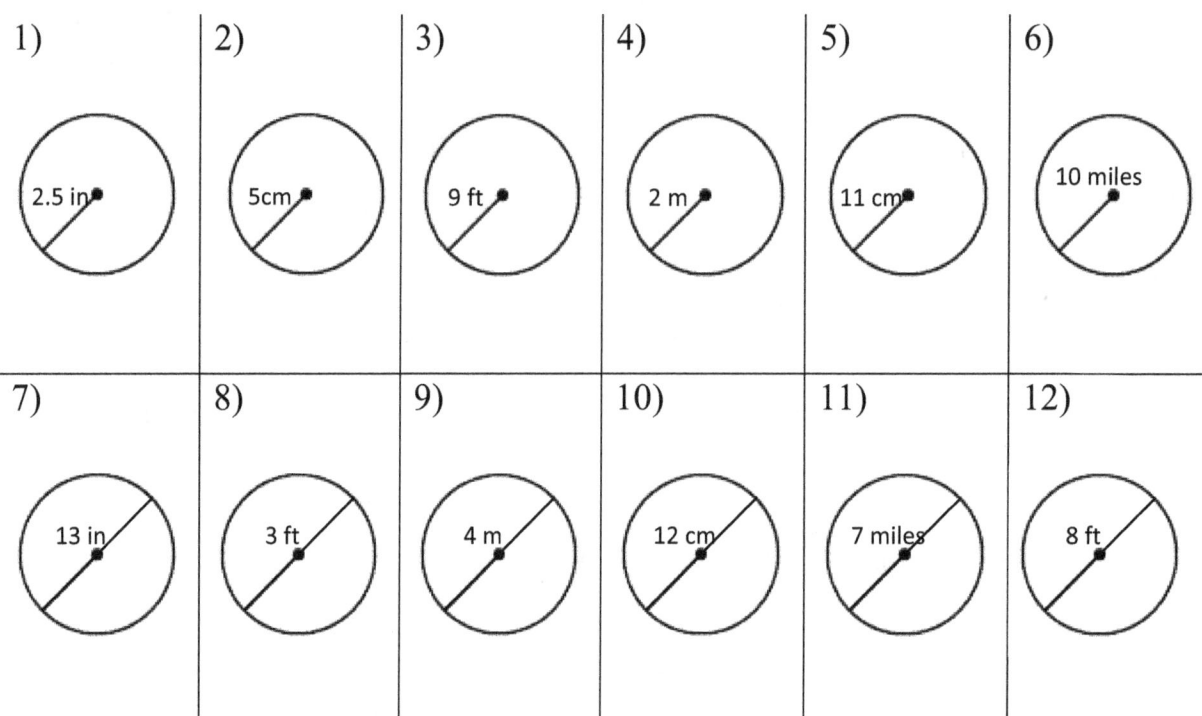

✏️ **Complete the table below.** ($\pi = 3.14$)

Circle No.	Radius	Diameter	Circumference	Area
1	1 in	2 in	6.28 in	3.14 in^2
2		10 m		
3				28.26 ft^2
4			47.1 mi	
5		11 km		
6	7 cm			
7		12 ft		
8				314 m^2
9			56.52 in	
10	4.5 ft			

WWW.MathNotion.Com

ACT Aspire Subject Test Mathematics Grade 7

Cubes

✏️ **Find the volume of each cube.**

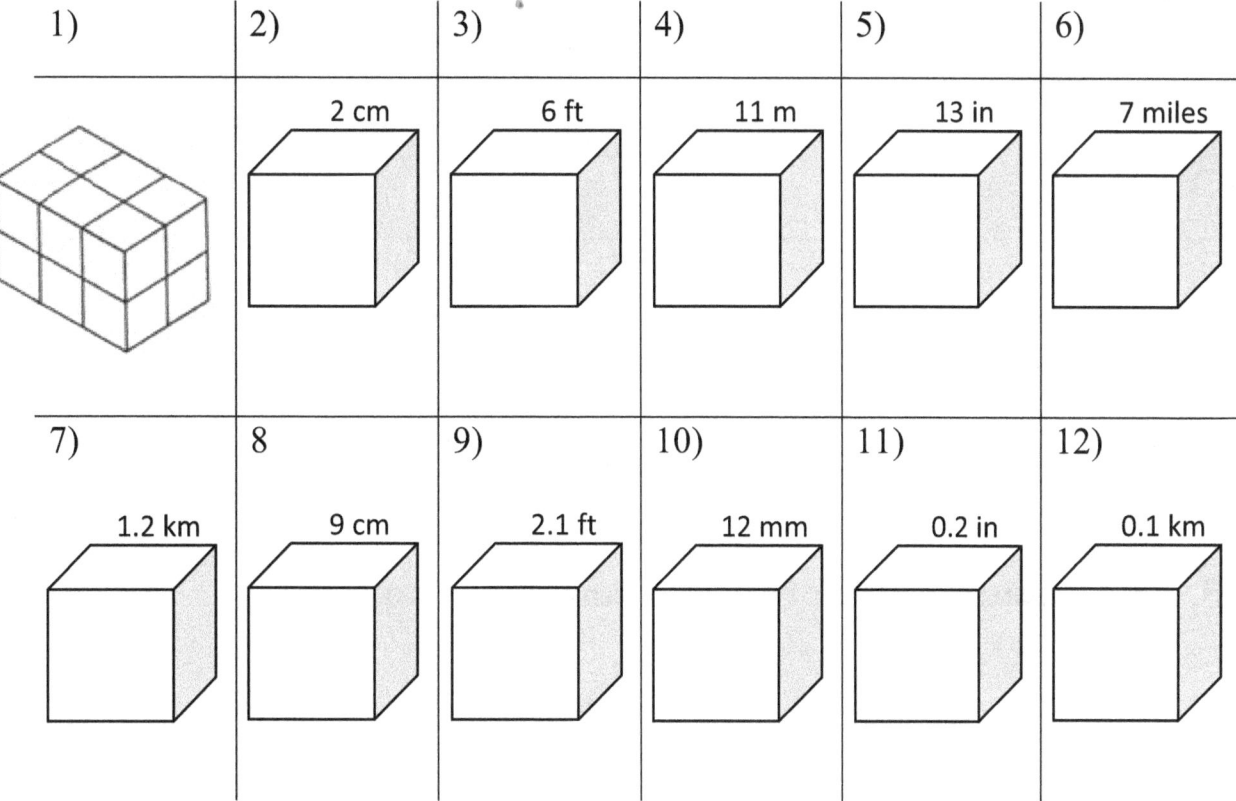

✏️ **Find the surface area of each cube.**

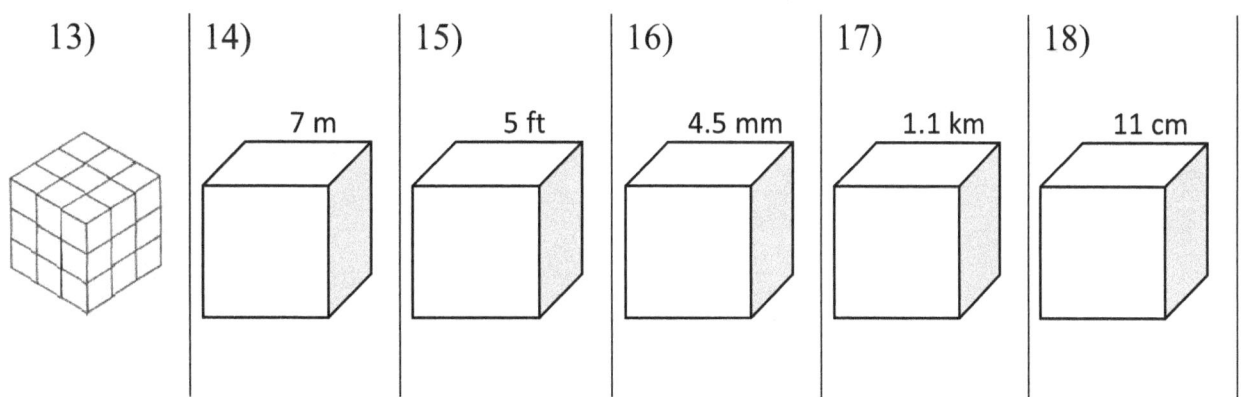

Rectangular Prism

✏️ **Find the volume of each Rectangular Prism.**

ACT Aspire Subject Test Mathematics Grade 7

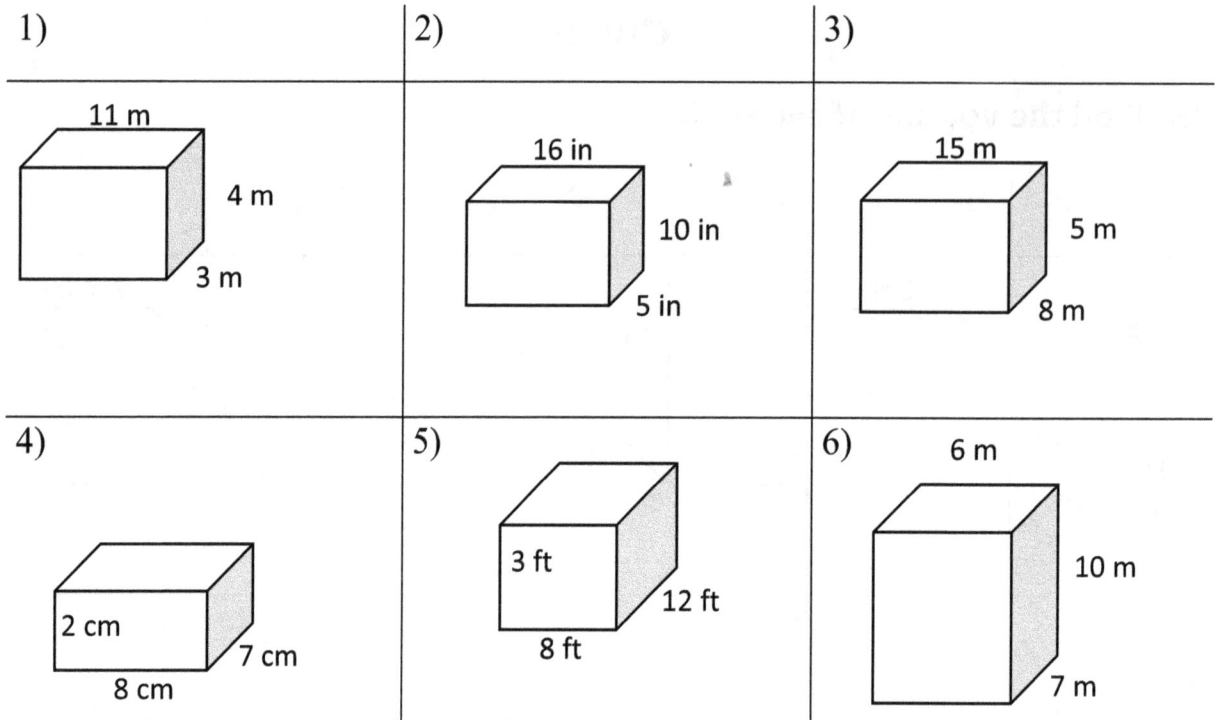

🖎 **Find the surface area of each Rectangular Prism.**

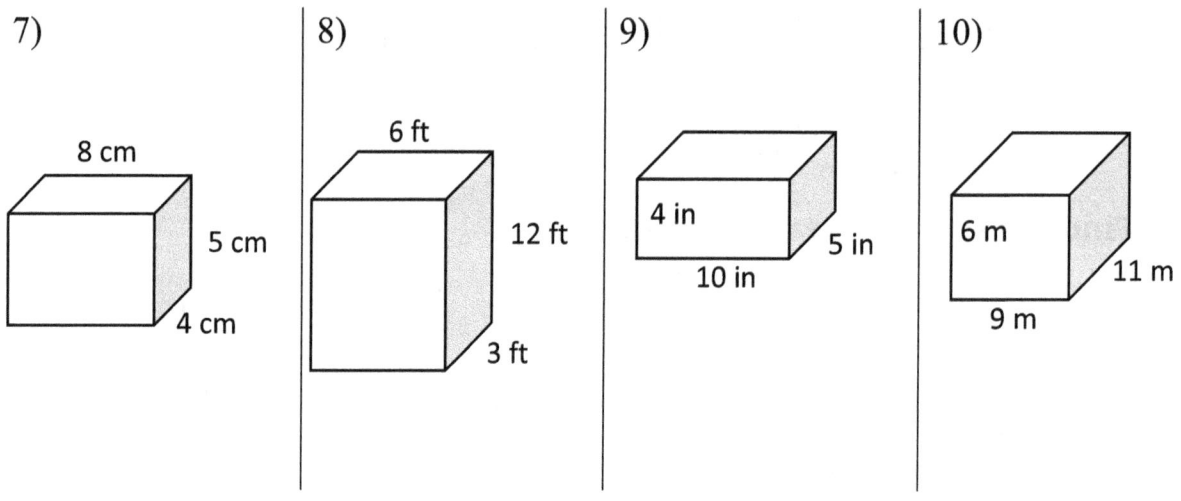

ACT Aspire Subject Test Mathematics Grade 7

Cylinder

✎ Find the volume of each Cylinder. Round your answer to the nearest tenth. ($\pi = 3.14$)

1)

2)

3)

4)

5)

6)
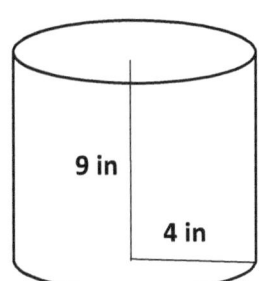

✎ Find the surface area of each Cylinder. ($\pi = 3.14$)

7)

8)

9)

10)
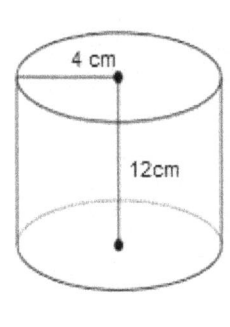

WWW.MathNotion.Com

ACT Aspire Subject Test Mathematics Grade 7

Pyramids and Cone

✎ Find the volume of each Pyramid and Cone. ($\pi = 3.14$)

1)

2)

3)

4)

5)

6)

✎ Find the surface area of each Pyramid and Cone. ($\pi = 3.14$)

7)

8)

9)

10)

WWW.MathNotion.Com   114

ACT Aspire Subject Test Mathematics Grade 7

Answers of Worksheets

Angles

1) 16°	4) 34°	7) 90°	10) 75°
2) 96°	5) 70°	8) 75°	11) 70°
3) 59°	6) 52°	9) 33°	

Pythagorean Relationship

1) No	5) Yes	9) 13	13) 15
2) Yes	6) No	10) 20	14) 30
3) No	7) Yes	11) 17	15) 36
4) Yes	8) Yes	12) 10	16) 12

Triangles

1) 60°	5) 45°	9) 54 square unites
2) 48°	6) 40°	10) 120 square unites
3) 55°	7) 45°	11) 90 square unites
4) 52°	8) 67°	12) 180 square unites

Polygons

1) 52 ft	5) 54 m	9) 30 m^2
2) 26 in	6) 25 cm	10) 300 in^2
3) 112 ft	7) 38 in	11) 160 km^2
4) 20 cm	8) 24 m	12) 49 in^2

Trapezoids

1) 50 cm^2	4) 60 cm^2	7) 36
2) 105 m^2	5) 80	8) 15
3) 39 ft^2	6) 24	

Calculate

1) 13 cm	2) 11 ft	3) 12 m	4) 25 ft

Circles

1) 19.63 in^2	5) 379.94 cm^2	9) 12.56 m^2
2) 78.5 cm^2	6) 314 $miles^2$	10) 113.04 cm^2
3) 254.34 ft^2	7) 132.67 in^2	11) 38.47 $miles^2$
4) 12.56 m^2	8) 7.07 ft^2	12) 50.24 ft^2

WWW.MathNotion.Com

ACT Aspire Subject Test Mathematics Grade 7

Circle No.	Radius	Diameter	Circumference	Area
1	1 in	2 in	6.28 in	3.14 in^2
2	5 m	10 m	31.4 m	78.5 m^2
3	3 ft	6 ft	18.84 ft	28.26 ft^2
4	7.5 miles	15 mi	47.1 mi	176.63 mi^2
5	5.5 km	11 km	34.54 km	94.99 km^2
6	7 cm	14 cm	43.96 cm	153.86 cm^2
7	6 ft	12 ft	37.68 feet	113.04 ft^2
8	10 m	20 m	62.8 m	314 m^2
9	9 in	18 in	56.52 in	254.34 in^2
10	4.5 ft	9 ft	28.26 ft	63.585 ft^2

Cubes

1) 12
2) 8 cm^3
3) 216 ft^3
4) 1,331 m^3
5) 2,197 in^3
6) 343 $miles^3$
7) 1.728 km^3
8) 729 cm^3
9) 9.261 ft^3
10) 1,728 mm^3
11) 0.008 in^3
12) 0.001 km^3
13) 27
14) 294 m^2
15) 150 ft^2
16) 121.5 mm^2
17) 7.26 km^2
18) 726 cm^2

Rectangular Prism

1) 132 m^3
2) 800 in^3
3) 600 m^3
4) 112 cm^3
5) 288 ft^3
6) 420 m^3
7) 184 cm^2
8) 252 ft^2
9) 220 in^2
10) 438 m^2

Cylinder

1) 1,004.8 m^3
2) 214.6 cm^3
3) 9,495.4 cm^3
4) 1.1 m^3
5) 588.8 m^3
6) 452.2 in^3
7) 188.4 m^2
8) 602.9 cm^2
9) 37.7 cm^2
10) 401.9 m^2

Pyramids and Cone

1) 1,600 yd^3
2) 1,050 yd^3
3) 1,617 in^3
4) 392.5 m^3
5) 3,014.4 m^3
6) 366.33 cm^3
7) 1,440 yd^2
8) 1,536 m^2
9) 678.24 in^2
10) 1,205.76 cm^2

Chapter 11:
Statistics and Probability

Topics that you will practice in this chapter:

- ✓ Mean and Median
- ✓ Mode and Range
- ✓ Histograms
- ✓ Stem–and–Leaf Plot
- ✓ Pie Graph
- ✓ Probability Problems

Mathematics is no more computation than typing is literature.
– John Allen Paulos

ACT Aspire Subject Test Mathematics Grade 7

Mean and Median

✎ **Find Mean and Median of the Given Data.**

1) 8, 7, 14, 4, 8

2) 14, 8, 25, 19, 16, 33, 11

3) 23, 18, 15, 12, 17

4) 34, 14, 10, 15, 6, 11

5) 10, 19, 6, 8, 32, 20, 17

6) 17, 26, 39, 69, 20, 6

7) 40, 38, 18, 11, 9, 2, 7, 32, 41

8) 24, 21, 31, 12, 33, 32, 22

9) 16, 14, 20, 41, 15, 20, 38, 4

10) 20, 20, 30, 18, 6, 28, 12, 46

11) 12, 7, 10, 11, 16, 22

12) 10, 29, 27, 12, 2, 15, 10, 3

✎ **Calculate.**

13) In a javelin throw competition, five athletics score 56, 34, 62, 23 and 19 meters. What are their Mean and Median? _____

14) Eva went to shop and bought 8 apples, 14 peaches, 6 bananas, 4 pineapples and 12 melons. What are the Mean and Median of her purchase? _____

15) Bob has 17 black pen, 19 red pen, 14 green pens, 20 blue pens and 5 boxes of yellow pens. If the Mean and Median are 19 respectively, what is the number of yellow pens in each box? _____

ACT Aspire Subject Test Mathematics Grade 7

Mode and Range

✎ **Find Mode and Rage of the Given Data.**

1) 4, 3, 7, 3, 3, 4
   Mode: _____   Range: _____

2) 18, 18, 24, 26, 18, 8, 14, 22
   Mode: _____   Range: _____

3) 8, 8, 8, 16, 19, 22, 20, 9, 13
   Mode: _____   Range: _____

4) 24, 24, 14, 28, 20, 18, 20, 24
   Mode: _____   Range: _____

5) 6, 21, 27, 24, 27, 27
   Mode: _____   Range: _____

6) 21, 8, 8, 7, 8, 12, 10, 22, 18, 13
   Mode: _____   Range: _____

7) 7, 4, 4, 6, 13, 13, 13, 0, 2, 2
   Mode: _____   Range: _____

8) 5, 8, 5, 14, 12, 14, 3, 5, 18
   Mode: _____   Range: _____

9) 7, 7, 7, 12, 7, 3, 8, 16, 3, 17
   Mode: _____   Range: _____

10) 15, 15, 19, 16, 4, 16, 10, 15
    Mode: _____   Range: _____

11) 6, 6, 5, 6, 42, 13, 19, 2
    Mode: _____   Range: _____

12) 8, 8, 9, 8, 9, 4, 34, 22
    Mode: _____   Range: _____

✎ **Calculate.**

13) A stationery sold 12 pencils, 56 red pens, 24 blue pens, 20 notebooks, 12 erasers, 21 rulers and 11 color pencils. What are the Mode and Range for the stationery sells?

    Mode: _____   Range: _____

14) In an English test, eight students score 10, 15, 15, 18 18, 16, 15 and 15. What are their Mode and Range? _____

15) What is the range of the first 6 even numbers greater than 8?
    _____

WWW.MathNotion.Com

ACT Aspire Subject Test Mathematics Grade 7

Times Series

✎ Use the following Graph to complete the table.

Day	Distance (km)
1	
2	

The following table shows the number of births in the US from 2007 to 2012 (in millions).

Year	Number of births (in millions)
2007	4.15
2008	3.70
2009	3.45
2010	3.20
2011	1.75
2012	2.98

Draw a Time Series for the table.

WWW.MathNotion.Com

Stem-and-Leaf Plot

✏ **Make stem ad leaf plots for the given data.**

1) 24, 26, 29, 20, 53, 27, 51, 55, 36, 21, 37, 30

   Stem | Leaf plot

2) 11, 59, 66, 14, 18, 19, 59, 65, 69, 61, 68, 65

   Stem | Leaf plot

3) 121, 55, 66, 54, 112, 128, 63, 125, 59, 123, 68, 119

   Stem | Leaf plot

4) 51, 32, 100, 56, 84, 36, 107, 56, 85, 39, 56, 106, 89

   Stem | Leaf plot

5) 33, 89, 19, 87, 81, 16, 11, 30, 86, 35, 17, 35, 13

   Stem | Leaf plot

6) 60, 92, 22, 25, 67, 93, 95, 62, 21, 64, 98, 29

   Stem | Leaf plot

Pie Graph

The circle graph below shows all Robert's expenses for last month. Robert spent $140 on his hobbies last month.

Answer following questions based on the Pie graph.

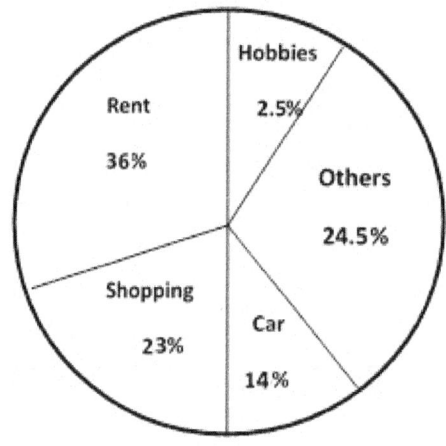

1) How much was Robert's total expenses last month? _____

2) How much did Robert spend on his car last month? _____

3) How much did Robert spend for shopping last month? _____

4) How much did Robert spend on his rent last month? _____

5) What fraction is Robert's expenses for his rent and car out of his total expenses last month? _____

ACT Aspire Subject Test Mathematics Grade 7

Probability Problems

✒ Calculate.

1) A number is chosen at random from 1 to 10. Find the probability of selecting number 6 or smaller numbers. _____

2) Bag A contains 18 red marbles and 6 green marbles. Bag B contains 16 black marbles and 8 orange marbles. What is the probability of selecting a green marble at random from bag A? What is the probability of selecting a black marble at random from Bag B? _____

3) A number is chosen at random from 1 to 20. What is the probability of selecting multiples of 4? _____

4) A card is chosen from a well-shuffled deck of 52 cards. What is the probability that the card will be a queen? _____

5) A number is chosen at random from 1 to 15. What is the probability of selecting a multiple of 3 or 5? _____

A spinner numbered 1–8, is spun once. What is the probability of spinning …?

6) an Odd number? _____   7) a multiple of 2? _____

8) a multiple of 5? _____   9) number 10? _____

ACT Aspire Subject Test Mathematics Grade 7

Answers of Worksheets

Mean and Median

1) Mean: 8.2, Median: 8
2) Mean: 18, Median: 16
3) Mean: 17, Median: 17
4) Mean: 15, Median: 12.5
5) Mean: 16, Median: 17
6) Mean: 29.5, Median: 23
7) Mean: 22, Median: 18
8) Mean: 25, Median: 24
9) Mean: 21, Median: 18
10) Mean: 22.5, Median: 20
11) Mean: 13, Median: 11.5
12) Mean: 13.5, Median: 11
13) Mean: 38.8, Median: 34
14) Mean: 8.8, Median: 8
15) 5

Mode and Range

1) Mode: 3, Range: 4
2) Mode: 18, Range: 18
3) Mode: 8, Range: 14
4) Mode: 24, Range: 14
5) Mode: 27, Range: 21
6) Mode: 8, Range: 15
7) Mode: 13, Range: 13
8) Mode: 5, Range: 15
9) Mode: 7, Range: 14
10) Mode: 15, Range: 15
11) Mode: 6, Range: 40
12) Mode: 8, Range: 30
13) Mode: 12, Range: 45
14) Mode: 15, Range: 8
15) 10

Time series

Day	Distance (km)
1	335
2	496
3	270
4	610
5	320
6	400

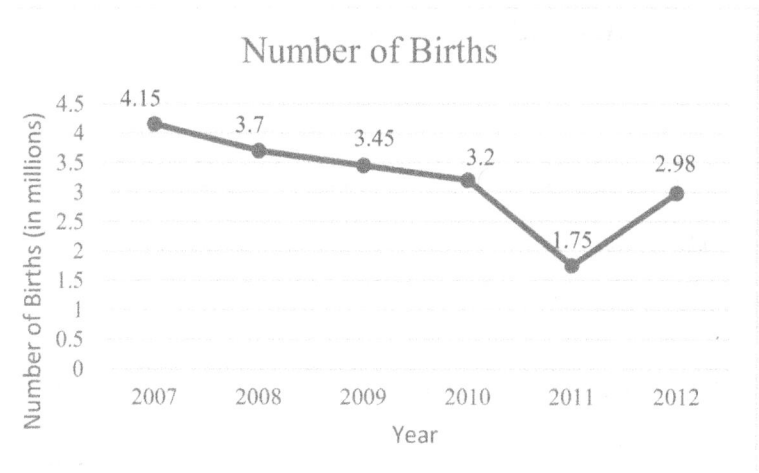

Stem–And–Leaf Plot

1)

Stem	leaf
2	0 1 4 6 7 9
3	0 6 7
5	1 3 5

2)

Stem	leaf
1	1 4 8 9
5	9 9
6	1 5 5 6 8 9

3)

Stem	leaf
5	4 5 9
6	3 6 8
11	2 9
12	1 3 5 8

WWW.MathNotion.Com

ACT Aspire Subject Test Mathematics Grade 7

4)

Stem	leaf
3	2 6 9
5	1 6 6 6
8	4 5 9
10	0 6 7

5)

Stem	leaf
1	1 3 6 7 9
3	0 3 5 5
8	1 6 7 9

6)

Stem	leaf
2	2 1 5 9
6	0 2 4 7
9	2 3 5 8

Pie Graph

1) $5,600

2) $784

3) $1,288

4) $2,016

5) $\frac{1}{2}$

Probability Problems

1) $\frac{3}{5}$

2) $\frac{1}{4}, \frac{2}{3}$

3) $\frac{1}{4}$

4) $\frac{1}{13}$

5) $\frac{7}{15}$

6) $\frac{1}{2}$

7) $\frac{1}{2}$

8) $\frac{1}{8}$

9) 0

ACT Aspire Subject Test Mathematics Grade 7

Chapter 12 : ACT Aspire Math Practice Tests

Time to Test

Time to refine your skill with a practice examination.

Take two practice Grade 7 ACT Aspire Math Tests to simulate the test day experience. After you've finished, score your test using the answer key.

Before You Start

- You'll need a pencil and scratch papers to take the test.

- For this practice test, don't time yourself. Spend time as much as you need.

- It's okay to guess. You won't lose any points if you're wrong.

- After you've finished the test, review the answer key to see where you went go.

Calculators are not permitted for students taking the ACT Aspire Mathematics Grade 7.

Good Luck!

ACT Aspire Subject Test Mathematics Grade 7

ACT Aspire Subject Test Mathematics Grade 7

ACT Aspire GRADE 7 MAHEMATICS REFRENCE MATERIALS

Linear Equations

Slope-intercept form $\qquad y = mx + b$

Constant of proportionality $\qquad k = \frac{y}{x}$

Circumference

Circle $\qquad C = 2\pi r \qquad$ or $\qquad C = \pi d$

Area

Triangle $\qquad A = \frac{1}{2}bh$

Rectangle or Parallelogram $\qquad A = bh$

Trapezoid $\qquad A = \frac{1}{2}h(b_1 + b_2)$

Circle $\qquad A = \pi r^2$

Volume

Prism or cylinder $\qquad V = Bh$

Pyramid or Cone $\qquad V = \frac{1}{3}Bh$

Additional Information

Pi $\qquad \pi = 3.14 \qquad$ or $\qquad \pi = \frac{22}{7}$

Distance $\qquad d = rt$

Simple interest $\qquad I = prt$

Compound interest $\qquad I = p(1 + r)^t$

ACT Aspire Subject Test Mathematics Grade 7

ACT Aspire Practice Test 1

Mathematics

GRADE 7

- **46 Questions:** 30 multiple choice questions and 16 grid-ins questions.
- **Total time for this test:** 60 Minutes.
- **You may use a calculator for this test.**

Released Month Year

ACT Aspire Subject Test Mathematics Grade 7

1) Which of the following points lies on the line 4x + 3y = 13?

   A. (1, 3)

   B. (–1, 4)

   C. (–3, 2)

   D. (3, 2)

2) 6 + 8 × (– 4) – [6 + 15 × 4] ÷ 6 =?

   Write your answer in the box below.

   ☐

3) The price of a car was $18,000 in 2014, $13,500 in 2015 and $10,125 in 2016.

   What is the rate of depreciation of the price of car per year?

   A. 15 %

   B. 28 %

   C. 26 %

   D. 25 %

4) What is the equivalent temperature of 572°F in Celsius?

   $$C = \frac{5}{9}(F - 32)$$

   A. 50

   B. 150

   C. 300

   D. 280

ACT Aspire Subject Test Mathematics Grade 7

5) The square of a number is $\frac{16}{36}$. What is the cube of that number?

  A. $\frac{8}{18}$

  B. $\frac{32}{72}$

  C. $\frac{64}{216}$

  D. $\frac{64}{144}$

6) What is the value of x in the following equation?

$$\frac{3}{5}x + \frac{1}{6} = \frac{1}{3}$$

Write your answer in the box below.

[ ]

7) What is the surface area of the cylinder below?

  A. 78π

  B. 154π

  C. 616π

  D. 308π

8) The average of five numbers is 44. If a sixth number 80 is added, then, what is the new average?

Write your answer in the box below.

[ ]

ACT Aspire Subject Test Mathematics Grade 7

9) Anita's trick–or–treat bag contains 16 pieces of chocolate, 12 suckers, 26 pieces of gum, 18 pieces of licorice. If she randomly pulls a piece of candy from her bag, what is the probability of her pulling out a piece of sucker?

A. $\frac{1}{4}$

B. $\frac{1}{6}$

C. $\frac{1}{12}$

D. $\frac{1}{16}$

10) Which of the following shows the numbers in descending order?

$$\frac{6}{5}, 0.67, 83\%, \frac{3}{8}$$

A. $83\%, 0.76, \frac{6}{5}, \frac{3}{8}$

B. $83\%, 0.76, \frac{3}{8}, \frac{6}{5}$

C. $0.76, 83\%, \frac{3}{8}, \frac{6}{5}$

D. $\frac{6}{5}, 83\%, 0.76, \frac{3}{8}$

11) The mean of 32 test scores was calculated as 75. But it turned out that one of the scores was misread as 70 but it was 54. What is the correct mean of the data?

Write your answer in the box below.

ACT Aspire Subject Test Mathematics Grade 7

12) The width of a box is one fifth of its length. The height of the box is one sixth of its width. If the length of the box is 90 cm, what is the volume of the box?

   A. 760 cm^3

   B. 1,880 cm^3

   C. 3,800 cm^3

   D. 4,960 cm^3

13) In five successive hours, a car travels 22 km, 24 km, 21 km, 29 km and 55 km. In the next five hours, it travels with an average speed of 36 km per hour. Find the total distance the car traveled in 10 hours.

   A. 331 km

   B. 620 km

   C. 210 km

   D. 151 km

14) In 1999, the average worker's income increased \$3,400 per year starting from \$43,000 annual salary. Which equation represents income greater than average? (I = income, x = number of years after 1999)

   A. I > 3,400 x + 43,000

   B. I > – 3,400 x + 43,000

   C. I < –3,400 x + 43,000

   D. I < 3,400 x – 43,000

ACT Aspire Subject Test Mathematics Grade 7

15) What is the slope of a line that is parallel to the line $8x - 4y = 16$?

Write your answer in the box below.

16) The perimeter of the trapezoid below is 29 cm. What is its area?

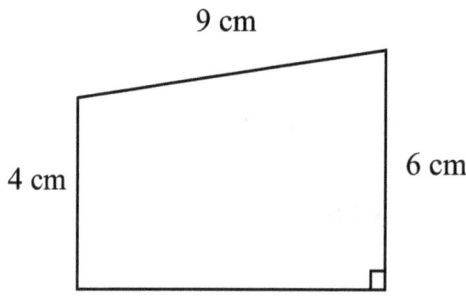

Write your answer in the box below.

17) Which of the following graphs represents the compound inequality $-4 \leq 5x - 9 < 21$?

A.

B.

C.

D.

ACT Aspire Subject Test Mathematics Grade 7

18) A football team had $45,000 to spend on supplies. The team spent $20,000 on new balls. New sport shoes cost $168 each. Which of the following inequalities represent how many new shoes the team can purchase?

A. $168x + 20,000 \leq 45,000$

B. $168x + 20,000 \geq 45,000$

C. $20,000x + 168 \leq 45,000$

D. $20,000x + 168 \geq 45,000$

19) What is the value of the expression $3(x - 4y) + (6 - x)^2$ when $x = 3$ and $y = 3$?

Write your answer in the box below.

[ ]

20) Two dice are thrown simultaneously, what is the probability of getting a sum of 6 or 10?

A. $\frac{5}{9}$

B. $\frac{4}{9}$

C. $\frac{1}{9}$

D. $\frac{2}{9}$

21) A swimming pool holds 2,100 cubic feet of water. The swimming pool is 30 feet long and 14 feet wide. How deep is the swimming pool?

Write your answer in the box below.

22) A bank is offering 4.3% simple interest on a savings account. If you deposit $6,200, how much interest will you earn in five years?

A. $199

B. $1,333

C. $666

D. $1,266

23) A card is drawn at random from a standard 52–card deck, what is the probability that the card is of aces?

A. $\frac{2}{13}$

B. $\frac{1}{13}$

C. $\frac{3}{52}$

D. $\frac{13}{52}$

24) Which graph corresponds to the following inequalities?

$$y \leq -3x + 6$$

$$-4x + 2y \leq 8$$

A.

B.

C.

D.

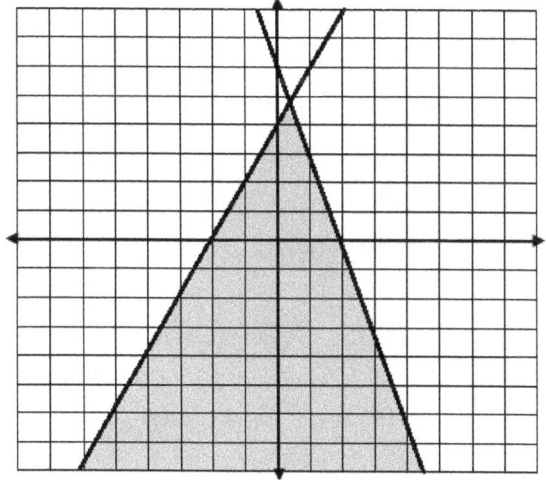

ACT Aspire Subject Test Mathematics Grade 7

25) How long does a 310–miles trip take moving at 50 miles per hour (mph)?

   A. 7 hours

   B. 6 hours and 20 minutes

   C. 6 hours and 15 minutes

   D. 6 hours and 12 minutes

26) 14 yards 5 feet and 9 inches equals to how many inches?

   Write your answer in the box below.

27) What is the area of a square whose diagonal is 16?

   Write your answer in the box below.

28) A shirt costing $700 is discounted 24%. After a month, the shirt is discounted another 24%. Which of the following expressions can be used to find the selling price of the shirt?

   A. (600) (0.76)

   B. (700) – 700 (0.24)

   C. (700) (0.24) – (400) (0.24)

   D. (700) (0.76) (0.76)

ACT Aspire Subject Test Mathematics Grade 7

29) Mr. Carlos family are choosing a menu for their reception. They have 7 choices of appetizers, 10 choices of entrees, 5 choices of cake. How many different menu combinations are possible for them to choose?

   A. 35

   B. 50

   C. 320

   D. 350

30) Seven one – foot rulers can be split among how many users to leave each with $\frac{1}{8}$ of a ruler?

   A. 7

   B. 15

   C. 56

   D. 42

31) The ratio of boys to girls in a school is 7: 5. If there are 720 students in a school, how many boys are in the school.

   Write your answer in the box below.

32) What is the volume of a box with the following dimensions?

   Height = 3 cm   Width = 6 cm   Length = 9 cm

   Write your answer in the box below.

33) What is the area of the shaded region?

A. 15 ft^2

B. 96 ft^2

C. 81 ft^2

D. 2,118 ft^2

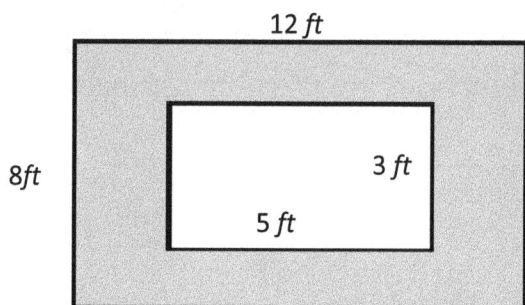

34) Mr. Jones saves $3,000 out of his monthly family income of $78,000. What fractional part of his income does he save?

A. $\frac{1}{6}$

B. $\frac{1}{16}$

C. $\frac{1}{26}$

D. $\frac{3}{26}$

35) The ratio of boys and girls in a class is 3: 4. If there are 63 students in the class, how many more boys should be enrolled to make the ratio 1: 1?

A. 9

B. 12

C. 36

D. 24

36) When a number is subtracted from 56 and the difference is divided by that number, the result is 7. What is the value of the number?

A. 2

B. 5

C. 7

D. 8

37) The radius of the following cylinder is 6 inches, and its height is 11 inches. What is the surface area of the cylinder?

A. 84 π cm²

B. 102 π cm²

C. 204 π cm²

D. 408 π cm²

38) How many tiles of 6 cm² is needed to cover a floor of dimension 4 cm by 54 cm?

Write your answer in the box below.

ACT Aspire Subject Test Mathematics Grade 7

39) In two successive years, the population of a town is increased by 16% and 35%. What percent of its population is increased after two years?

A. 46.6

B. 66.6

C. 56.6

D. 54.6

40) In a school, the ratio of number of boys to girls is 6: 7. If the number of boys is 240, what is the total number of students in the school?

Write your answer in the box below.

☐

41) What is the value of x in the following equation? $9^x = 729$

Write your answer in the box below.

☐

42) What is value of $-19 - (-54)$?

Write your answer in the box below.

☐

43) What number is 14 more than 18% of 150?

A. 18

B. 26

C. 41

D. 27

ACT Aspire Subject Test Mathematics Grade 7

44) If $\frac{4x}{9} = 80$, then $\frac{5x}{9} = ?$

   A. 50

   B. 75

   C. 75

   D. 100

45) If 32 % of a class are girls, and 30 % of girls play tennis, what percent of the class play tennis?

   A. 96 %

   B. 1.96%

   C. 9.6 %

   D. 0.96 %

46) If x = lowest common multiple of 24 and 36 then $\frac{x}{4} - 7$ equal to?

   A. 25

   B. 18

   C. 11

   D. 13

STOP

This is the End of Test 1.

"You may check your work on this Test if you still have time".

ACT Aspire Practice Test 2

Mathematics

GRADE 7

- **46 Questions:** 30 multiple choice questions and 16 grid-ins questions.
- **Total time for this test:** 60 Minutes.
- **You may use a calculator for this test.**

Released *Month Year*

ACT Aspire Subject Test Mathematics Grade 7

1) If $x = -8$, which equation is true?

   A. $x(3x + 3) = -44$

   B. $7(6 - x) = 98$

   C. $6(3x + 10) = 89$

   D. $12x - 20 = -66$

2) In a bag of small balls $\frac{1}{6}$ are black, $\frac{1}{4}$ are white, $\frac{1}{12}$ are red and the remaining 8 blue. How many balls are white?

   Write your answer in the box below.

3) A boat sails 18 miles south and then 24 miles east. How far is the boat from its start point?

   A. 35

   B. 22

   C. 15

   D. 30

4) The area of a circle is 16π. What is the circumference of the circle?

   A. 4π

   B. 8π

   C. 16π

   D. 32π

ACT Aspire Subject Test Mathematics Grade 7

5) Sophia purchased a sofa for $517. The sofa is regularly priced at $550. What was the percent discount Sophia received on the sofa?

   A. 1.4%

   B. 6%

   C. 94%

   D. 1.04%

6) The score of Elise was one fourth as that of Mia and the score of Stella was twice that of Mia. If the score of Stella was 72, what is the score of Elise?

   Write your answer in the box below.

7) A bag contains 18 balls: seven green, three black, four blue, one brown, two red and one white. If 12 balls are removed from the bag at random, what is the probability that a brown ball has been removed?

   A. $\frac{2}{3}$

   B. $\frac{1}{12}$

   C. $\frac{1}{15}$

   D. $\frac{1}{18}$

8) A rope weighs 375 grams per meter of length. What is the weight in kilograms of 11.4 meters of this rope? (1 kilograms = 1000 grams)

   A. 0.04275

   B. 0.4275

   C. 4.275

   D. 4,275

9) What is the area of the shaded region?

   A. 91π in^2

   B. 50π in^2

   C. 51π in^2

   D. 92π in^2

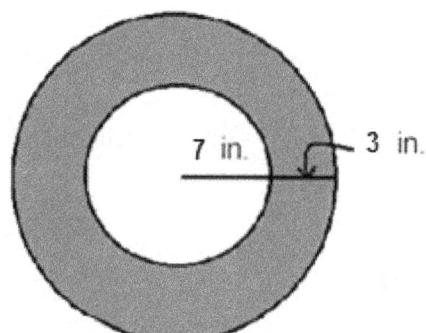

10) A pizza cut into 6 parts. Elise and her sister Etta ordered two pizzas. Elise ate $\frac{2}{3}$ of her pizza and Etta ate $\frac{1}{3}$ of her pizza. What part of the two pizzas was left?

   A. $\frac{7}{12}$

   B. $\frac{1}{4}$

   C. $\frac{1}{2}$

   D. $\frac{5}{12}$

11) Simplify $3y^4(2x^5y)^4 =$

Write your answer in the box below.

12) Right triangle ABC has two shorter legs of lengths 21 cm (AB) and 28 cm (AC). What is the length of the third side (BC)?

A. 42 cm

B. 40 cm

C. 38 cm

D. 35 cm

13) The marked price of a computer is D dollar. Its price decreased by 34% in January and later increased by 15 % in February. What is the final price of the computer in D dollar?

A. 0.81 D

B. 0.66 D

C. 0.76 D

D. 1.16 D

14) $[4 \times (-28) + 26] - (32) + [7 \times 15] \div 3 =?$

Write your answer in the box below.

ACT Aspire Subject Test Mathematics Grade 7

15) A $160 shirt now selling for $80 is discounted by what percent?

   A. 120 %

   B. 50 %

   C. 90 %

   D. 80 %

16) From last year, the price of gasoline has increased from $3.25 per gallon to $3.965 per gallon. The new price is what percent of the original price?

   A. 22 %

   B. 62 %

   C. 162 %

   D. 122 %

17) What is the median of these numbers? 17, 41, 21, 63, 28, 10, 56

   Write your answer in the box below.

18) The following trapezoid are similar. What is the value of x ?

   A. 18

   B. 15

   C. 16

   D. 28

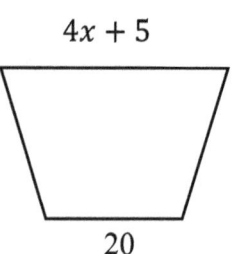

ACT Aspire Subject Test Mathematics Grade 7

19) If 75% of a class are girls, and 24% of girls play tennis, what fraction of the class play tennis?

   A. 36.20%

   B. 18.30%

   C. 18%

   D. 36%

20) A chemical solution contains 12% alcohol. If there is 96 ml of alcohol, what is the volume of the solution?

   Write your answer in the box below.

   [ ]

21) The price of a laptop is decreased by 10% to $ 630. What is its original price?

   A. 750

   B. 700

   C. 500

   D. 500

22) What is the perimeter of a square that has an area of 375.21 feet?

   Write your answer in the box below.

   [ ]

ACT Aspire Subject Test Mathematics Grade 7

23) Last week 18,000 fans attended a football match. This week five times as many bought tickets, but one ninth of them cancelled their tickets. How many are attending this week?

   A. 90,000

   B. 80,000

   C. 89,000

   D. 88,000

24) Two times the price of a laptop is equal to six times the price of a computer. If the price of laptop is $800 more than the computer, what is the price of the computer?

Write your answer in the box below.

☐

25) Jason is 15 miles ahead of Joe running at 4 miles per hour and Joe is running at the speed of 7 miles per hour. How long does it take Joe to catch Jason?

   A. 11 hours

   B. 5 hours

   C. 3 hours

   D. 9 hours

26) Jason needs an 79% average in his writing class to pass. On his first 4 exams, he earned scores of 75%, 88%, 91%, and 83%. What is the minimum score Jason can earn on his fifth and final test to pass?

Write your answer in the box below.

27) 60 students took an exam and 18 of them failed. What percent of the students passed the exam?

A. 30%

B. 18%

C. 70%

D. 55%

28) What is the volume of the following square pyramid?

Write your answer in the box below.

29) A bank is offering 3.6% simple interest on a savings account. If you deposit $42,000, how much interest will you earn in three years?

   A. $4,536

   B. $3,044

   C. $2,680

   D. $2,860

30) 31 less than twice a positive integer is 73. What is the integer?

   Write your answer in the box below.

   $\boxed{}$

31) Which of the following points lies on the line $-3x + 4y = 6$?

   A. (4, 2)

   B. (−6, −3)

   C. (−2, 5)

   D. (3, 1)

32) How many possible outfit combinations come from six shirts, two slacks, and seven ties?

   Write your answer in the box below.

   $\boxed{}$

ACT Aspire Subject Test Mathematics Grade 7

33) An angle is equal to one eighth of its supplement. What is the measure of that angle?

   A. 15

   B. 20

   C. 36

   D. 45

34) John traveled 360 km in 20 hours and Alice traveled 720 km in 10 hours. What is the ratio of the average speed of John to average speed of Alice?

   A. 1: 6

   B. 1: 4

   C. 5: 9

   D. 1: 8

35) In a party, 9 soft drinks are required for every 30 guests. If there are 510 guests, how many soft drinks is required?

   Write your answer in the box below.

   ☐

36) 4.8 is what percent of 32?

   A. 4.6

   B. 28

   C. 14

   D. 15

ACT Aspire Subject Test Mathematics Grade 7

37) You can buy 7 cans of green beans at a supermarket for $14.70. How much does it cost to buy 26 cans of green beans?

   A. $48.6

   B. $54.60

   C. $44.00

   D. $42.8

38) Joe scored 37 out of 42 marks in Algebra, 40 out of 68 marks in science and 18 out of 24 marks in mathematics. In which subject his percentage of marks is best?

   A. Algebra

   B. Mathematics and Algebra

   C. Algebra and Science

   D. Science

39) What is the length of BC in the following figure if AB = 5, DF = 8 and BD = 26?

   A. 5

   B. 10

   C. 14

   D. 18

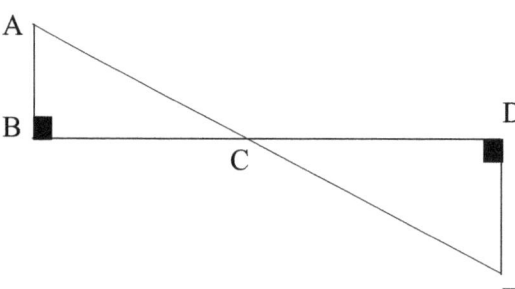

ACT Aspire Subject Test Mathematics Grade 7

40) Robert is preparing to run a marathon. He runs $5\frac{1}{4}$ miles on Saturday and four times that many on Monday and Wednesday. Robert wants to run a total of 40 miles this week. How many more miles does he need to run?

Write your answer in the box below.

41) A shaft rotates 200 times in 5 seconds. How many times does it rotate in 25 seconds?

A. 1,210

B. 1,000

C. 1,300

D. 600

42) 14.520 ÷ 0.006?

Write your answer in the box below.

43) If 64% of A is 16% of B, then B is what percent of A?

A. 4%

B. 40%

C. 400%

D. 440%

ACT Aspire Subject Test Mathematics Grade 7

44) Find $\frac{2}{5}$ of $\frac{5}{6}$ of $\frac{1}{7}$ of 630?

   A. 25

   B. 75

   C. 30

   D. 40

45) If $x \leq a$ is the solution of $7 + 9x \leq 52$, what is the value of a?

   A. $9x$

   B. 5

   C. -6

   D. $6x$

46) Solve for x: $6 + 3x + 12\left(\frac{x}{3}\right) = 4x + 30$

   Write your answer in the box below.

STOP

This is the End of Test 2.

"You may check your work on this Test if you still have time".

ACT Aspire Subject Test Mathematics Grade 7

Answers and Explanations

ACT Aspire Practice Tests

Answer Key

❋ Now, it's time to review your results to see where you went wrong and what areas you need to improve!

ACT Aspire - Mathematics

Practice Test - 1						Practice Test - 2					
1	A	16	50	31	420	1	B	16	D	31	B
2	−37	17	D	32	162	2	4	17	28	32	84
3	D	18	A	33	C	3	D	18	B	33	B
4	C	19	−18	34	C	4	B	19	C	34	B
5	C	20	D	35	A	5	B	20	800	35	153
6	$\frac{5}{18}$	21	5	36	C	6	9	21	B	36	D
7	D	22	B	37	C	7	D	22	75.6	37	B
8	50	23	B	38	36	8	C	23	B	38	A
9	B	24	D	39	C	9	C	24	400	39	B
10	D	25	D	40	520	10	C	25	B	40	$13\frac{3}{4}$
11	74.5	26	573	41	3	11	$48\,x^{20}y^8$	26	58	41	B
12	D	27	128	42	35	12	D	27	C	42	2,420
13	A	28	D	43	C	13	C	28	1,014	43	C
14	A	29	D	44	D	14	−83	29	A	44	C
15	2	30	C	45	C	15	B	30	52	45	B
				46	C					46	8

WWW.MathNotion.Com

ACT Aspire Subject Test Mathematics Grade 7

Practice Test 1

Answers and Explanations

1) Answer: A.

Input (1, 3) in the 4x + 3y = 13 formula instead of x and y. So, we have: 4(1) + 3(3) = 13 $\Rightarrow$ 4 + 9 = 13

2) Answer: – 37.

Use PEMDAS (order of operation):

6 + 8 × (–4)– [6 + 15 × 4] ÷ 6 = 6 + 8 × (–4) – [6 + 60] ÷ 6 = 6 + 8 × (–4) – [66] ÷ 6 = 6 + (–32) – 11= 6 + (–32) – 11 = –26 – 11 = –37

3) Answer: D.

Use this formula: Percent of Change = $\frac{New\ Value - Old\ Value}{Old\ Value} \times 100\ \%$

$\frac{13,500-18,000}{18,000} \times 100\ \% = -25\ \%$ and $\frac{10,125-13,500}{13,500} \times 100\ \% = -25\ \%$

4) Answer: C.

Plug in 572 for F and then solve for C.

C = $\frac{5}{9}$ (F – 32) $\Rightarrow$ C = $\frac{5}{9}$ (572 – 32) $\Rightarrow$ C = $\frac{5}{9}$ (540) = 300

5) Answer: C.

The square of a number is $\frac{16}{36}$, then the number is the square root of $\frac{16}{36}$: $\sqrt{\frac{16}{36}} = \frac{4}{6}$

The cube of the number is: $(\frac{4}{6})^3 = \frac{64}{216}$

6) Answer: $\frac{5}{18}$.

$\frac{3}{5}x + \frac{1}{6} = \frac{1}{3} \Rightarrow \frac{3}{5}x = \frac{1}{6} \Rightarrow x = \frac{5}{3} \times \frac{1}{6} \Rightarrow x = \frac{5}{18}$

7) Answer: D.

Surface Area of a cylinder = 2πr (r + h),

The radius of the cylinder is 7 (14 ÷ 2) inches, and its height is 16 inches. Therefore,

Surface Area of a cylinder = 2π (7) (7+ 15) = 308 π

ACT Aspire Subject Test Mathematics Grade 7

8) Answer: 50.

Solve for the sum of five numbers.

average = $\frac{\text{sum of terms}}{\text{number of terms}}$ ⇒ $44 = \frac{\text{sum of 5 numbers}}{5}$ ⇒ sum of 5 numbers = $44 \times 5 = 220$

The sum of 5 numbers is 220. If a sixth number 80 is added, then the sum of 6 numbers is $220 + 80 = 300$ ⇒ average = $\frac{\text{sum of terms}}{\text{number of terms}} = \frac{300}{6} = 50$

9) Answer: B.

Probability = $\frac{\text{number of desired outcomes}}{\text{number of total outcomes}} = \frac{12}{16+12+26+18} = \frac{12}{72} = \frac{1}{6}$

10) Answer: D.

Change the numbers to decimal and then compare.

$\frac{3}{8} = 0.375$; 0.67; $83\% = 0.83$; $\frac{6}{5} = 1.2$.

Therefore $\frac{6}{5} > 83\% > 0.67 > \frac{3}{8}$

11) Answer: 74.5.

average (mean) = $\frac{\text{sum of terms}}{\text{number of terms}}$ ⇒ $75 = \frac{\text{sum of terms}}{32}$ ⇒ sum = $75 \times 32 = 2,400$

The difference of 70 and 54 is 16. Therefore, 16 should be subtracted from the sum.

$2,400 - 16 = 2,384$

mean = $\frac{\text{sum of terms}}{\text{number of terms}}$ ⇒ mean = $\frac{2,384}{32} = 74.5$

12) Answer: D.

If the length of the box is 90, then the width of the box is one fifth of it, 18, and the height of the box is 3 (one sixth of the width). The volume of the box is:

$V = l \times w \times h = (90) \times (18) \times (3) = 4,860$

13) Answer: A.

Add the first 5 numbers. $22 + 24 + 21 + 29 + 55 = 151$

To find the distance traveled in the next 5 hours, multiply the average by number of hours.

Distance = Average × Rate = $36 \times 5 = 180$

Add both numbers. $151 + 180 = 331$

ACT Aspire Subject Test Mathematics Grade 7

14) Answer: A.

Let x be the number of years. Therefore, $3,400 per year equals $3,400x$.

starting from $43,000 annual salary means you should add that amount to $3,400x$.

Income more than that is: $I > 3,400x + 43,000$.

15) Answer: 2.

The equation of a line in slope intercept form is: $y = mx + b$

Solve for y.

$8x - 4y = 16 \Rightarrow -4y = 16 - 8x \Rightarrow y = (16 - 8x) \div (-4) \Rightarrow y = 2x - 4$

The slope of this line is 2. Parallel lines have same slopes.

16) Answer: 50.

The perimeter of the trapezoid is 29 cm.

Therefore, the missing side (high) is $= 29 - 9 - 6 - 4 = 10$

Area of a trapezoid: $A = \frac{1}{2} h (b_1 + b_2) = \frac{1}{2} (10)(4 + 6) = 50$

17) Answer: D.

Solve for x.

$-4 \leq 5x - 9 < 21 \Rightarrow$ (add 9 all sides) $-4 + 9 \leq 5x - 9 + 9 < 21 + 9 \Rightarrow 5 \leq 5x < 30 \Rightarrow$ (divide all sides by 5) $1 \leq x < 6$

x is between 1 and 6. Choice C represent this inequality.

18) Answer: A.

Let x be the number of new shoes the team can purchase. Therefore, the team can purchase $168x$.

The team had $45,000 and spent $20,000. Now the team can spend on new shoes $25,000 at most.

Now, write the inequality: $168x + 20,000 \leq 45,000$

19) Answer: -18

When $x = 3$ and $y = 3$,

Substitute: $3(x - 4y) + (6 - x)^2 = 3(3 - 4(3)) + (6 - (3))^2$
$= 3(3 - 12) + (6 - 3)^2 = -27 + 9 = -18$

ACT Aspire Subject Test Mathematics Grade 7

20) Answer: D.

The options to get sum of 6: (1 & 5) and (5 & 1), (2 & 4), (4 & 2) and (3, 3), so we have 5 options.

The options to get sum of 10: (4 & 6), (6 & 4), (5 & 5) we have 3 options.

To get the sum of 6 or 10 for two dice, we have 8 options: $5 + 3 = 8$

Since, we have $6 \times 6 = 36$ total options, the probability of getting a sum of 6 and 10 is 8 out of 36 or $\frac{8}{36} = \frac{2}{9}$.

21) Answer: 5.

Use formula of rectangle prism volume.

V = (length) (width) (height) $\Rightarrow$ 2,100 = (30) (14) (height) $\Rightarrow$ height = 2,100 ÷ 420 = 5

22) Answer: B.

Use simple interest formula:

$I = prt$ (I = interest, p = principal, r = rate, t = time)

$I = (6,200)(0.043)(5) = 1,333$

23) Answer: B.

The probability of choosing aces is $\frac{4}{52} = \frac{1}{13}$.

24) Answer: D.

For each option, choose a point in the solution part and check it on both inequalities.

$$y \leq -3x + 6$$
$$-4x + 2y \leq 8$$

A. Point (1, 7) is in the solution section. Let's check the point in both inequalities.

   $7 \leq -3(1) + 6$, That's not true

B. Let's choose this point (–3, 2); $2 \leq -2(3) + 6$, That's true!

   $-4(-3) + 2(2) \leq 8 \Rightarrow 14 \leq 8$, That's not true

C. Let's choose this point (3, 3); $3 \leq -3(3) + 6$, That's not true

D. Let's choose this point (0, 0); $0 \leq -3(0) + 6$, That's true!

   $-4(0) + 2(0) \leq 8 \Rightarrow 0 \leq 8$, it works (this point works in both)

ACT Aspire Subject Test Mathematics Grade 7

25) Answer: D.

Use distance formula:

Distance = Rate × time ⇒ 310 = 50 × T,

divide both sides by 40. $\frac{310}{50}$ = T ⇒ T = 6.2 hours.

Change hours to minutes for the decimal part. 0.2 hours = 0.2 × 60 = 12 minutes.

26) Answer: 573.

14 × 36 + 5 × 12 + 9 = 573

27) Answer: 128.

The diagonal of the square is 16. Let x be the side.

Use Pythagorean Theorem: $a^2 + b^2 = c^2$

$x^2 + x^2 = 16^2 \Rightarrow 2x^2 = 256 \Rightarrow x^2 = 128 \Rightarrow x = \sqrt{128}$

The area of the square is: $\sqrt{128} \times \sqrt{128} = 128$

28) Answer: D.

To find the discount, multiply the number by (100% – rate of discount).

Therefore, for the first discount we get: (700) (100% – 24%) = (700) (0.76) = 420

For the next 24 % discount: (700) (0.76) (0.76)

29) Answer: D.

To find the number of possible outfit combinations, multiply number of options for each factor: 7 × 10 × 5 = 350

30) Answer: C.

$7 \div \frac{1}{8} = 56$

31) Answer: 420.

The ratio of boy to girls is 7:5. Therefore, there are 7 boys out of 12 students. To find the answer, first divide the total number of students by 12, then multiply the result by 7.

720 ÷ 12 = 60 ⇒ 60 × 7 = 420

32) Answer: 162.

Volume of a box = length × width × height = 9 × 6 × 3 = 162

ACT Aspire Subject Test Mathematics Grade 7

33) Answer: C.

Use the area of rectangle formula ($s = a \times b$).

To find area of the shaded region subtract the smaller rectangle from bigger rectangle.

$S_1 - S_2 = (12ft \times 8t) - (5ft \times 3ft) \Rightarrow S_1 - S_2 = 96ft^2 - 15ft^2 = 81ft^2$

34) Answer: C.

3,000 out of 78,000 equals to $\frac{3,000}{78,000} = \frac{3}{78} = \frac{1}{26}$

35) Answer: A.

The ratio of boy to girls is 3:4. Therefore, there are 3 boys out of 7 students.

To find the answer, first divide the total number of students by 7, then multiply the result by 3. $63 \div 7 = 9 \Rightarrow 9 \times 3 = 27$

There are 27 boys and 36 (63 – 27) girls. So, 9 more boys should be enrolled to make the ratio 1:1.

36) Answer: C.

Let the number be x. Then: $\frac{56-x}{x} = 7 \rightarrow 7x = 56 - x \rightarrow 8x = 56 \rightarrow x = 7$

37) Answer: C.

Surface Area of a cylinder = 2πr (r + h),

The radius of the cylinder is 6 inches, and its height is 11 inches. π is about 3.14. Then:

Surface Area of a cylinder = 2 (π) (6) (6 + 11) = 204 π = 640.56

38) Answer: 36.

The area of the floor is: 4 cm × 54 cm = 216 cm²

The number of tiles needed = 216 ÷ 6 = 36

39) Answer: C.

The population is increased by 16% and 35%. 16% increase changes the population to 116% of original population.

For the second increase, multiply the result by 135%.

(1.16) × (1.35) = 1.566 = 156.6%

56.6 percent of the population is increased after two years.

ACT Aspire Subject Test Mathematics Grade 7

40) Answer: 520.

The ratio of boy to girls is 6:7. Therefore, there are 6 boys out of 13 students. To find the answer, first divide the number of boys by 6, then multiply the result by 13.

$240 \div 6 = 40 \Rightarrow 40 \times 13 = 520$

41) Answer: 3.

$81 = 9^2 \to 9^x = (3^2)^x = 3^{2x}$

$729 = 3^6 \to (3)^{2x} = 3^6 \to 2x = 6 \to x = 3$

42) Answer: 35.

$-19 - (-54) = -19 + 54 = 54 - 19 = 35$

43) Answer: C.

18% of 150 = $\frac{18}{100} \times 150 = 27$; Let x be the number then, $x = 27 + 14 = 41$

44) Answer: D.

If $\frac{4x}{9} = 80$. then $4x = 720 \to x = 180$

$\frac{5x}{9} = \frac{5 \times 180}{9} = \frac{900}{9} = 100$

45) Answer: C.

The percent of girls playing tennis is: $32\% \times 30\% = 0.32 \times 0.30 = 0.096 = 9.6\%$

46) Answer: C.

Prime factorizing of 24 = $2 \times 2 \times 2 \times 3$

Prime factorizing of 36 = $2 \times 2 \times 3 \times 3$

x = LCM = $2 \times 2 \times 2 \times 3 \times 3 = 72$

$\frac{72}{4} - 7 = 18 - 7 = 11$

ACT Aspire Subject Test Mathematics Grade 7

Practice Test 2
Answers and Explanations

1) Answer: B.

Only option B is correct. Other options don't work in the equation.

$7(6 − (−8)) = 98$

2) Answer: 4.

Let x be the number of balls. Then: $\frac{1}{6}x + \frac{1}{4}x + \frac{1}{12}x + 8 = x$

$(\frac{1}{6} + \frac{1}{4} + \frac{1}{12})x + 8 = x$

$(\frac{1}{2})x + 8 = x \Rightarrow x = 16$

In the bag of small balls $\frac{1}{4}$ are white, then: $\frac{16}{4} = 4$

There are 4 white balls in the bag.

3) Answer: D.

Use the information provided in the question to draw the shape.

Use Pythagorean Theorem: $a^2 + b^2 = c^2$

$18^2 + 24^2 = c^2 \Rightarrow 324 + 576 = c^2 \Rightarrow 900 = c^2 \Rightarrow c = 30$

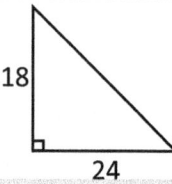

4) Answer: B.

Use the formula of areas of circles.

Area $= \pi r^2 \Rightarrow 16\pi = \pi r^2 \Rightarrow 16 = r^2 \Rightarrow r = 4$

Radius of the circle is 4. Now, use the circumference formula:

Circumference $= 2\pi r = 2\pi (4) = 8\pi$

5) Answer: B.

The question is this: 517 is what percent of 550?

Use percent formula: part $= \frac{\text{percent}}{100} \times$ whole

$517 = \frac{\text{percent}}{100} \times 550 \Rightarrow 517 = \frac{\text{percent} \times 550}{100}$

$\Rightarrow 51,700 =$ percent $\times 550 \Rightarrow$ percent $= \frac{51,700}{550} = 94$

517 is 94 % of 550. Therefore, the discount is: 100% − 94% = 6%

WWW.MathNotion.Com

ACT Aspire Subject Test Mathematics Grade 7

6) Answer: 9.

If the score of Stella was 72, then the score of Mia is 36. Since the score of Elise was one fourth as that of Mia, therefore, the score of Elise is 9.

7) Answer: D.

If 12 balls are removed from the bag at random, there will be one ball in the bag. The probability of choosing a brown ball is 1 out of 18. Therefore, the probability of not choosing a brown ball is 12 out of 18 and the probability of having not a brown ball after removing 12 balls is the same.

8) Answer: C.

The weight of 11.4 meters of this rope is: 11.4 × 375 g = 4,275 g

1kg = 1000 g, therefore, 4,275 g ÷ 1000 = 4.275 kg

9) Answer: C.

To find the area of the shaded region, find the difference of the area of two circles. (S_1: the area of bigger circle. S_2: the area of the smaller circle)

Use the area of circle formula. $S = \pi r^2$

$S1 - S2 = \pi(7+3)^2 - \pi(7in)^2 \Rightarrow S1 - S2 = \pi 100 in^2 - \pi 49\, in^2 \Rightarrow$
$S1 - S2 = 51\pi\, in^2$

10) Answer: C.

Elise ate $\frac{2}{3}$ of 6 parts of her pizza that it means 4 parts out of 6 parts ($\frac{2}{3}$ of 6 parts = x ⇒ x = 4) and left 2 parts.

Ella ate $\frac{1}{3}$ of 6 parts of her pizza that it means 2 parts out of 6 parts ($\frac{1}{3}$ of 6 parts = x ⇒ x = 2) and left 4 parts.

Therefore, they ate (4 + 2) parts out of (6 + 6) parts of their pizza and left (4 + 2) parts out of (6 + 6) parts of their pizza. It means: $\frac{6}{12}$.

After simplification we have: $\frac{1}{2}$

11) Answer: $48x^{20}y^8$.

Simplify: $3y^4(2x^5y)^4 = 3y^4(16x^{20}y^4) = 48x^{20}y^8$

ACT Aspire Subject Test Mathematics Grade 7

12) Answer: D.

Use Pythagorean Theorem: $a^2 + b^2 = c^2$

$21^2 + 28^2 = c^2 \Rightarrow 441 + 784 = c^2 \Rightarrow c^2 = 1,225 \Rightarrow c = 35$

13) Answer: C.

To find the discount, multiply the number by (100% − rate of discount).

Therefore, for the first discount we get: (D) (100% − 34%) = (D) (0.66) = 0.66 D

For increase of 15 %: (0.66 D) (100% + 15%) = (0.66 D) (1.15) = 0.759 D = 76% of D

14) Answer: −83.

Use PEMDAS (order of operation):

$[4 \times (-28) + 26] - (32) + [7 \times 15] \div 3 =$

$= [-112 + 26] - 32 + [105] \div 3 = [-86] - 32 + 35 = -83$

15) Answer: B.

Use the formula for Percent of Change: $\frac{New\ Value - Old\ Value}{Old\ Value} \times 100\%$

$\frac{80-160}{160} \times 100\% = -50\%$

(negative sign here means that the new price is less than old price).

16) Answer: D.

The question is this: 3.965 is what percent of 3.25?

Use percent formula: $part = \frac{percent}{100} \times whole$

$3.965 = \frac{percent}{100} \times 3.25 \Rightarrow 3.965 = \frac{percent \times 3.25}{100} \Rightarrow 396.5 = percent \times 3.25$

$\Rightarrow percent = \frac{396.5}{3.25} = 122$

17) Answer: 28.

Write the numbers in order: 10, 17, 21, 28, 41, 56, 63

Since we have 7 numbers (7 is odd), then the median is the number in the middle, which is 28.

18) Answer: B.

Write the ratio and solve for x.

$\frac{78}{24} = \frac{4x+5}{20} \Rightarrow 24(4x+5) = 78 \times 20 \Rightarrow 4x+5 = 1,560 \div 24 \Rightarrow 4x+5 = 65 \Rightarrow x = 15$

ACT Aspire Subject Test Mathematics Grade 7

19) Answer: C.

Let x be the number of students in the class.

75% of x = girls

24 % of girls = tennis player; Find 24% of 75%. Then:

$24\% \text{ of } 75\% = 0.24 \times 0.75 = 0.18 = 18\%$

20) Answer: 800.

12% of the volume of the solution is alcohol. Let x be the volume of the solution.

Then: 12% of x = 96 ml $\Rightarrow 0.12\ x = 96 \Rightarrow x = 96 \div 0.12 = 800$

21) Answer: B.

Let x be the original price.

If the price of a laptop is decreased by 10% to $630, then:

$90\ \% \text{ of } x = 630 \Rightarrow 0.90x = 630 \Rightarrow x = 630 \div 0.90 = 700$

22) Answer: 75.6.

Use the area of square formula: $S = a^2 \Rightarrow 375.21 = a^2 \Rightarrow a = 18.9$

One side of the square is 18.9 feet. Use the perimeter of square formula.

$P = 4a \Rightarrow P = 4(18.9) \Rightarrow P = 75.6$

23) Answer: B.

Five times of 18,000 is 90,000. One ninth of them cancelled their tickets.

One ninth of 90,000 equal 10,000 ($\frac{1}{9} \times 90,000 = 10,000$).

80,000 (90,000 – 10,000 = 80,000) fans are attending this week

24) Answer: $400.

Let L be the price of laptop and C be the price of computer.

$2(L) = 6(C)$ and $L = \$800 + C$

Therefore, $2(\$800 + C) = 6C \Rightarrow \$1,600 + 2C = 6C \Rightarrow 4C = \$1,600 \Rightarrow C = \$400$

25) Answer: B.

The distance between Jason and Joe is 15 miles. Jason running at 4 miles per hour and Joe is running at the speed of 7 miles per hour. Therefore, every hour the distance is 3 miles less. $15 \div 3 = 5$

ACT Aspire Subject Test Mathematics Grade 7

26) Answer: 58.

Jason needs an 79% average to pass for five exams. Therefore, the sum of 5 exams must be at least $5 \times 79 = 395$

The sum of 4 exams is: $75 + 88 + 91 + 83 = 337$.

The minimum score Jason can earn on his fifth and final test to pass is: $395 - 337 = 58$

27) Answer: C.

The failing rate is 18 out of $60 = \frac{18}{60}$

Change the fraction to percent: $\frac{18}{60} \times 100\% = 30\%$

30 percent of students failed. Therefore, 70 percent of students passed the exam.

28) Answer: $1,014 m^3$.

Use the volume of square pyramid formula.

$V = \frac{1}{3} a^2 h \Rightarrow V = \frac{1}{3}(13m)^2 \times 18m \Rightarrow V = 1,014$ m³

29) Answer: A.

Use simple interest formula: $I = prt$ (I = interest, p = principal, r = rate, t = time)

$I = (42,000)(0.036)(3) = 4,536$

30) Answer: 52.

Let x be the integer. Then: $2x - 31 = 73$

Add 31 both sides: $2x = 104$. Divide both sides by 2: $x = 52$

31) Answer: B.

Input the points instead of x and y in the formula. Only option B works in the equation.

$-3x + 4y = 6$

$-3(-6) + 4(-3) = 6$

32) Answer: 84.

To find the number of possible outfit combinations, multiply number of options for each factor: $6 \times 2 \times 7 = 84$

33) Answer: B.

The sum of supplement angles is 180. Let x be that angle. Therefore, $x + 8x = 180$

$9x = 180$, divide both sides by 9: $x = 20$

ACT Aspire Subject Test Mathematics Grade 7

34) Answer: B.

The average speed of John is: 360 ÷ 20 = 18 km

The average speed of Alice is: 720 ÷ 10 = 72 km

Write the ratio and simplify: 18: 72 ⇒ 1: 4.

35) Answer: 153.

Let x be the number of soft drinks for 510 guests. It's needed to have a proportional ratio to find x.

$\frac{9 \text{ soft drinks}}{30 \text{ guests}} = \frac{x}{510 \text{ guests}} \Rightarrow x = \frac{510 \times 9}{30} \Rightarrow x = 153$

36) Answer: D.

x % of 32 is 4.8, then: x % 32 = 4.8 ⇒ 0.32 x = 4.8 ⇒ x = 4.8 ÷ 0.32 = 15

37) Answer: B.

Let x be the number of cans. Write a proportion and solve for x.

$\frac{7 \text{ cans}}{\$14.70} = \frac{26 \text{ cans}}{x} \Rightarrow x = \frac{14.70 \times 26}{7} \Rightarrow x = 54.6$

38) Answer: A.

Compare each score:

In Algebra Joe scored 37 out of 42 in Algebra that it means 88.09% of total mark: $\frac{37}{42} = \frac{x}{100} \Rightarrow x = 88.09$

Joe scored 40 out of 68 in science that it means 58.82% of total mark. $\frac{40}{68} = \frac{x}{100} \Rightarrow x = 58.82$

Joe scored 18 out of 24 in mathematic that it means 75% of total mark: $\frac{18}{24} = \frac{x}{100} \Rightarrow x = 75$

Therefore, his score in Algebra is higher than his other scores.

39) Answer: B.

Two triangles ∆ABC and ∆CD are similar. Then:

$\frac{AB}{DF} = \frac{BC}{CD} \to \frac{5}{8} = \frac{x}{26-x} \to 130 - 5x = 8x \to 13x = 130 \to x = 10$

40) Answer: $13\frac{3}{4}$ miles.

Robert runs $5\frac{1}{4}$ miles on Saturday and $4(5\frac{1}{4})$ miles on Monday and Wednesday.

Robert wants to run a total of 40 miles this week.

ACT Aspire Subject Test Mathematics Grade 7

Therefore, subtract $5\frac{1}{4} + 4(5\frac{1}{4})$ from 40. $40 - (5\frac{1}{4} + 4(5\frac{1}{4})) = 40 - 26\frac{1}{4} = 13\frac{3}{4}$ miles

41) Answer: B.

Number of rotates in 25 second = $\frac{200 \times 25}{5} = 1,000$

42) Answer: 2,420.

$14.520 \div 0.006 = \frac{\frac{14,520}{1,000}}{\frac{6}{1.000}} = \frac{14,520}{6} = 2,420$

43) Answer: C.

Write the equation and solve for B:

$0.64A = 0.16B$, divide both sides by 0.16, then you will have $\frac{0.64}{0.16}A = B$, therefore:

$B = 4A$, and B is 4 times of A or it's 400% of A.

44) Answer: C.

$\frac{1}{7}$ of 630 = $\frac{1}{7} \times 630 = 90$

$\frac{5}{6}$ of 90 = $\frac{5}{6} \times 90 = 75$

$\frac{2}{5}$ of 75 = $\frac{2}{5} \times 75 = 30$

45) Answer: B.

$7 + 9x \leq 52 \rightarrow 9x \leq 52 - 7 \rightarrow 9x \leq 45 \rightarrow x \leq \frac{45}{9} \rightarrow x \leq 5$ Then: $a = 5$

46) Answer: 8.

$6 + 3x + 12\left(\frac{x}{3}\right) = 4x + 30 \rightarrow 6 + 3x + 4x = 4x + 30 \rightarrow x = 8$

"End"

www.ingramcontent.com/pod-product-compliance
Lightning Source LLC
Chambersburg PA
CBHW080439110426
42743CB00016B/3215